P9-CSB-272

Hedges [illegible]
July. Octr. 1834.

Rubus Fruticosus
Bramble. Blackberry Bush
Icosandria Polygynia

THE FLOWERS OF MAY

Introduction and botanical notes by Richard Mabey

COLLINS & BROWN

First published in Great Britain in 1990 by Collins & Brown Limited
Mercury House, 195 Knightsbridge, London SW7 1RE

Copyright © Introduction Richard Mabey 1990
Copyright © Other texts Alphabet & Image Ltd 1990
Designed and produced by Alphabet & Image Ltd,
Sherborne, Dorset DT9 3LU
ISBN 1-85585-030-3

All rights reserved. No part of this publication may be reproduced, stored in a retrieval system, or transmitted in any form or by any means electronic, mechanical, photocopying, recording or otherwise, without the prior written permission of the copyright owner.

A CIP catalogue record for this book is available from the British Library.

Illustration on the half-title page:
Honeysuckle Lonicera Periclymenum, Common Honeysuckle
Dawlish, June 1831 No. 385, Vol 2

Illustration on the title page:
Bramble Rubus fruticosus, Blackberry Bush
Hedges, Breamore, July, October 1834 No. 262, Vol 2

Typesetting by August Filmsetting, Haydock, St Helens
Printed and bound in Hong Kong by Regent Publishing Services Ltd

Introduction

What was it about flower-painting that made it such a popular activity with Victorian women? Its image as a fashionable social grace, a genteel form of occupational therapy, hardly explains the zeal and diligence which women like Caroline May brought to their chosen pastime. At first sight she seems to fit perfectly into the stereotype of the artistic dabbler. She was born in 1809, the eldest daughter of a comfortably-off West Country vicar, and remained a spinster till her death in 1874. The first paintings she thought worth keeping were dated 1831 (see page 1 and Plate 74) — the same year that the distinguished botanist John Loudon gave the fashion his blessing in the *Gardener's Magazine*. 'To be able to draw Flowers botanically,' he wrote, 'is one of the most useful accomplishments of your ladies of leisure, living in the country.'

Yet it is hard to see Caroline May as one of those 'ladies of leisure'. Over a period of forty years she painted in most of the southern counties of England, and built up a portfolio of nearly a thousand finished pieces (covering close on half the native British flora). They show an exquisite attention to fine detail, to the subtle differences that could be seen in individual plants of the same species, and to the changes brought on by weather and ageing.

The captions penned at the foot of each picture confirm her enterprising spirit. Alongside the date and the plant's name, she almost always adds the location where it was found, and there seems to be nowhere that was too wet, wild or derelict for her. She found an opium poppy in a Cornish potato field, wallflowers on the battlements of Porchester Castle, water chickweed in a ditch in Surrey, shining cranesbill in a timber yard, and a lobelia 'apparently wild by the edge of a sewer near Sherbourn [sic], Dorset.'

She combed bogs and marshes and tracts of woodland with a relish that one doesn't expect to find in a middle-aged, well-bred Victorian lady; and anyone who has ever hunted for plants will sense in her a kindred spirit, and recognize that insatiable taste for tracking choice species down — buttonholing them, so to speak, in their 'proper' places. She had the wit to paint box from a specimen on Box Hill and found her pasque flower on the same Cambridgeshire chalk-hills where John Ray noted it two hundred years before.

By the standards of that time she travelled quite extensively about the British Isles, and lived at various times in Hampshire, Surrey and Cornwall (more details of her life and family history are given on pages 8 to 14). In Surrey especially she explored — and painted flowers in — a wide circle of country around her home, stretching to Windsor Great Park in the north-west and the Thames-side meadows between Walton and Kingston to the east. This pattern, of plant-hunting journeys radiating out from her home, recurred wherever she was living. From her base in Hampshire she haunted the Isle of Wight and the coastal marshland around Milton; in Cornwall she visited Bodmin Moor, the heathlands of the Lizard Peninsula and Padstow.

There were other trips as well, to the Channel Islands (four times), to Scotland, Shropshire, north Norfolk and the flower-rich dunes of Dawlish Warren in Devon. No wonder, perhaps, that in a lifetime's searching she was able to come across plants that are star rarities by our depleted standards, and for which no official county records exist even from her own time. She recorded, and accurately painted, downy woundwort *Stachys germanica* (now confined to a few sites in north Oxfordshire) in Surrey, and what appears to be the military orchid

Pennyroyal *Mentha pulegium*, painted in Chertsey, October 1846. This is now regarded as an endangered species in Britain, though it still has a locality in Surrey.

(now down to just two sites in Britain) in her home parish in Cornwall.

But her finest pictures are from the countryside in which she was brought up, the mosaic of ancient woods and water meadows round Breamore, with the bogs and heaths of the New Forest little more than a mile away to the east and the Wiltshire downlands to the north. She would go out in a small horse-drawn cart, possibly accompanied by a man, for protection and for gathering in the more intractable sites. Her plant identification was so accurate that she must have possessed — and been familiar with — some of the better botanical handbooks of her day. She may have had access to Smith and Sowerby's *English Botany*, whose twelve-volume second edition was published between 1831 and 1846; William Baxter's *British Phaenogamous Botany* (six vols, 1833-43) and the work of another enterprising lady, Anne Pratt, whose *Grasses, sedges and ferns of Great Britain* was published in 1860, just before Caroline May began her own work on these groups. There were very few local floras available at this time, but after 1863 Caroline could have had her confidence bolstered by one of the botanical best-sellers of the day, Margaret Plues's *Rambles in Search of Wild Flowers*. Miss Plues had stirring words of encouragement for those doughty women like Caroline May who studied plants in their proper setting — out in the field: 'Fanny might be seen, with an expression of lively interest on her countenance, climbing the cliffs, penetrating the woods, and exploring the salt marshes. Her mother thanked God for the renewed health which tinged her cheek and gave elasticity to her step, and she gladly procured stronger boots and dresses of firm texture in which she might ramble and climb at her own free will without fear of detriment. She made no objection even to the large flower-press, which would have been thought too uncouth for many a less elegant drawing-room; nay, she quite loved the rough machine as a means of procuring health and interest to her daughter.'

But there is no question of any 'borrowing' from these books. Caroline was working at levels of detail some way beyond those available in published work, and seemed to have a natural instinct for tracking plants down. She had a particular knack for finding odd colour varieties and sorts of common plants like yarrow (Plate 42) and cowslip (Plate 49). In the vicarage orchard in South Petherwyn in 1852 she found a double variety of lady's smock, and added a spray of it to a painting of the common single variety she had done at Breamore in 1834 (Plate 4).

This facility for integrating new elements into existing paintings, sometimes after long intervals (there is a gap of thirty-three years between the painting of the flowers and fruits of wild madder (Plate 38), is an example of her skill in composing a plant on the page. It is almost as if she had deliberately, but discreetly, left space for additions in the original picture.

She worked in watercolour, using varnish occasionally to emphasize the sheen on subjects like blackberries (frontispiece, page 2). She used very fine brushes to paint stamens and individual leaf-hairs, and was one of the few amateur painters of her day to bother with the grass family. Her brilliantly sharp portraits of the sedges (see Plates 78, 79 and 80) are arguably the most outstanding in this selection.

Common Centaury *Centaurium erythraea*. Started in Dawlish in June 1831, this was one of the earliest watercolours kept by Caroline May. But she did not finish it until nearly forty years later, when she added the blush variety, found at a site in Cornwall in 1870. Thus it marks both the beginning and the conclusion of her flora.

But she was just as adept at broad, bold washes. Her greater periwinkle (Plate 51) catches perfectly the simplicity of structure of this species, down to the propeller-like twists in the big petals. Her umbellifers (notoriously difficult subjects) have both delicacy and strength, and the hemlock water dropwort (Plate 33), especially, straddles the page without the slightest sacrifice of detail in the tiny white flowers. She had the deftest of touches in balancing a subject — turning three scabious flower-heads in different directions (Plate 39); dropping a single, emphasized leaf below and in front of a clump of dog violets (Plate 7); crossing the long seed-pods of yellow horned poppy (Plate 3) behind a single flower, as if they were a kind of heraldic device. With a few twists in the leaves she could even turn a nettle into a beguiling portrait (Plate 68).

This imaginative exploration of flowers during the painting echoes the enjoyment she clearly had in searching for them. She had, at both stages, an unerring instinct for *good* plants — for the seasonal, the quirky (see her henbane, Plate 55), the classically beautiful, the challengingly commonplace.

The poet Andrew Young, in one of his pleasantly discursive essays on wild flowers, makes the distinction between botanists and what he calls 'botanophiles'. Botanists are preoccupied with classifying plants, and unravelling how they work. Botanophiles are drawn towards flowers for their romance and associations, and especially to the excitement of hunting them down. Caroline May was both, and an incomparable artist to boot.

Caroline May mounted her flower paintings in five thick volumes, and at her death these were bequeathed to a nephew. They have been handed down through the family, and are currently owned by a third generation nephew, the Reverend John Tyler. Each volume is arranged in accordance with nineteenth-century Linnaean taxonomy, and there is no correspondence between the sequence of originals in the albums and the date of the painting. The selection of paintings reproduced here follows her own sequence, with roughly equal numbers taken from each volume.

Amongst the pages of paintings are contemporary maps of the places where Caroline painted most frequently. Readers who are able to visit these sites will be able to see the changes in habitat (and flora) for themselves.

Richard Mabey

Detail from an undated composite page of mosses.

Caroline May

In one respect Caroline May owed a lot to good luck: the talent she possessed exactly matched what was permitted by the period in which she lived and her situation. She was born shortly before George III finally succumbed to madness, and died when Victoria had been thirteen years a widow. Her whole maturity was passed in possibly the most repressive age that women have ever endured. In addition her entire life was spent in the shadow of Anglican clergymen — first her father and then her brother. Except for the fourteen years spent in Chertsey, her home was a vicarage, and the Victorian vicarage was not a destiny to be taken lightly. However, Gilbert White had set a precedent for the study of natural history, even in clergy households (to be sure, botany was preferable for the ladies, since one never knew where natural history might lead), and painting in watercolours had been rendered quite acceptable as an accomplishment for young ladies soon after Reeves began to manufacture dry blocks of watercolour paint in the mid-eighteenth century. Elegant watercolour boxes resembling tea-caddies, with neat drawers for phials and brushes, could be found in the most elevated households. Caroline was able to paint while the Brontë sisters kept their writing a secret, and while Charlotte writhed under the critics' reflections on her 'coarse' and 'unwomanly' portrayal of the emotions; she was painting while Florence Nightingale ate her heart out in frustration before being allowed to take up nursing in her thirties. Luckily, for a lady in the mid-nineteenth century, botany was irreproachable, and watercolours unassailable.

Surprisingly, Caroline's talent completely transcends the stultifying restrictions of her situation. Her paintings have a self-sufficiency and observed truth that makes them an effective contrast to the flaccid and affected flower-paintings of later and, at present, better-known ladies.

Caroline's family background illustrates the typical vicissitudes of the second half of the eighteenth century, and the Napoleonic Wars. Her great-grandfather, Thomas May, had made a considerable fortune in the Portuguese wine trade. He might well have lost it all again, had it not been for a mysterious incident which ever after ranked first in importance among the family's anecdotes. One day he received an urgent letter from his wife in England imploring him to come home at once; he disposed of his immediate business in Lisbon and set out post-haste to England. Rejoining his wife, he asked why she had sent for him. 'But I never did!' she said, and when shown the letter could only protest, 'The writing is mine, but I never wrote it.' It was some days before news of the terrible earthquake which had devastated Lisbon in 1755 followed Thomas to England. A portrait of Mrs May was commissioned in which she was represented holding the letter which had saved her husband, but his financial losses were still considerable. Nine years later he lost a further £12,000 worth of property when the Custom House at Lisbon was entirely destroyed by fire, and his son Joseph had much trouble in trying to satisfy the creditors of the family firm.

Joseph (known later in the family as 'Good Joseph', to distinguish him from his son 'Weak Joseph', and his grandson 'Wicked Joseph') married Mary, the daughter of his father's partner. They returned to England in 1775, and four years later Joseph inherited what was described (with decorous eighteenth-century restraint) as a 'considerable property' from a Mr Gardiner, a friend of his father's. This enabled the couple to purchase Hale Park in Hampshire, a 'pleasant country residence,

Hale Park, Hampshire, a lithograph by T. Boys. The home of Caroline May's uncle Joseph, Hale Park was the background for her childhood.

where he might spend the autumn of his days at a distance from the Metropolis.' It was and is an elegant and spacious house commanding magnificent prospects over the valley of the River Avon.

Here they proceeded to raise a numerous family: eleven children were born between 1766 and 1785. The second son, born on September 8th, 1772, was Thomas Charles May (known among his family as 'Handsome Thomas'). At the age of 24 he took Holy Orders, and his Aunt Margaret's diary recalls that 'on October 2nd he read Prayers to a large congregation (including his two grandmothers, his father, his mother, his aunt, three brothers and three sisters) and administered the Sacrament for the first time to his Grandmother May, aged 86, and his other Relations .... It was a most aweful scene and the last time that my dear Mother and Brother received it.' Not all clergy aspired to such solemnity; Thomas' sister, Maria Amelia, eloped with the Rev. Jeremiah Awdrey three years later, leaving the family's house between one and two in the morning. 'The event gave the family great concern,' observed her aunt. Meanwhile Thomas himself had married (in far more respectable circumstances) Mary Mawbey, one of the daughters of Sir Joseph Mawbey. Although this Baronet's fortunes fluctuated alarmingly in the risky financial atmosphere of the Napoleonic Wars, he was later able to present Thomas to the living of Chertsey, which he added to the family livings of Breamore and Hale — the holding of more than one living at the same time being still permitted in the Church of England at that period. But Mary died after only one year of marriage, perhaps in childbirth. 'Tom suffers greatly,' wrote his brother Joseph, with moving simplicity, to brother John in Lisbon.

But generally Hale Park provided a happy and exhilarating background for the family. 'We had a great many partys and my dear brother Joseph went with his family to Balls,' notes Aunt Margaret. And when Thomas performed the marriage ceremony for his elder brother and Miss Frances Stert, 'a month was spent in partys for the Bride and Bridegroom and to see Stone Henge, Wilton etc. and there was a great deal of music and other amusements.' It was at any rate sufficient to distract the diarist from her health (she was prone to suffer from a mysterious complaint known as St Anthony's Fire

Handsome Thomas, Caroline's father.

which required frequent 'Bleeding and Blistering'). Even when there were no social distractions the ladies could always ride out in the 3,000 acre estate on their Bouros (donkeys).

After seven years as a widower, Thomas married Rebecca Gibbons, daughter of Sir William Gibbons of Stanwell Place, Middlesex, and a friend of his first wife. Rebecca was descended from Bishop Trelawney, on whose behalf (according to the unofficial Cornish National anthem) twenty thousand Cornishmen threatened to march on London when James II locked him up with six other bishops in the Tower. The Bishop had given his daughter a small ring with 'Rebecca' on it, and this ring had been passed down the generations to Rebecca. Another, very different, seventeenth-century ancestor of hers was the woodcarver Grinling Gibbons.

Thomas and Rebecca were very happy together but in January 1809, when she was expecting their first child, they had a narrow escape. They had travelled from Hale to Breamore Rectory in Mrs May's carriage, going by Salisbury because of the deep snow and intense frost. On the way back the servants seem to have attempted to ford the river Avon where a bridge had been destroyed by floods; the two horses died instantly in the freezing storm-waters, and the two men-servants were lucky not to share the same fate. Mercifully Rebecca escaped and she was safely brought to bed of a fine girl two months later; the baby was christened Caroline Rebecca. The nursery at Breamore Rectory was later enlarged by the birth of a son, Henry, in 1814, and three more daughters, Louisa (1816), Elizabeth (1820) and Anna Maria (1822).

During Caroline's childhood and youth her father's elder brother, 'Weak Joseph', with his wife Frances, lived at Hale with their five children, Joseph, Charles, Fanny, Harriet and Emily. There was constant social intercourse between the families, and the cousins were much of an age. The younger Joseph was a wild young man, however, and his reckless gambling brought the Hale family into serious financial difficulties. He managed to mortgage Hale secretly for £60,000 to pay some of his debts, but he represented it as entailed solely to him after his father's death, whereas in fact, if he died without male issue, it was entailed to pass successively to his brother Charles, his Uncle Thomas and his cousin, Henry. When the creditors fore-

Breamore Church

closed the mortgage, his relatives were faced with the choice of breaking the entail and resigning all rights to Hale, or seeing Joseph committed to prison. The situation was exacerbated by the fact that 'Handsome Thomas' had put all the money he could into the property to try to save it, but he and Henry saw it as a matter of the family's honour and agreed to break the entail. The elder Joseph died at Versailles in 1830; his unregenerate eldest son tried to live at Hale for a while, let it to tenants, and finally sold it in 1836. He withdrew to France taking all the family pictures and miniatures as well as every small item of furniture that he could contrive to remove. The Hale Mays settled at Versailles, where it was possible to live more economically than in England. 'Wicked Joseph's' mother died there at the age of 89 in 1861.

This was a grim period for both branches of the family in that Caroline's father died in 1837, after holding the Hampshire livings for forty-two years, and the Chertsey living for thirty-two years. He left his widow with four unmarried daughters, their ages ranging from 28 to 15, and his son Henry, recently elected to a Fellowship of New College, Oxford. The family had to leave Breamore and Hale with their many happy and sad associations and move to Chertsey, where they rented a house called Ruxbury, found for them by friends, for £52 a year, purchasing the fittings from the previous tenant. Though variously described in early particulars as a 'villa' and 'the cottage of Ruxbury', it was, in fact, a pleasant house with two parlours and six bedchambers, and with stabling and outbuildings set in grounds of just under 7 acres. The Mays' nearest neighbour was the widow of Charles James Fox, occupying St Ann's Hill a quarter of a mile away. To the south of Ruxbury, Lady Hotham (an Admiral's widow) lived at Silverlands, and Mrs May was soon persuaded that the situation would provide some opportunities for social intercourse.

But for Caroline herself the outlook seemed much more bleak. She was approaching the age of thirty, and lived in a society which treated the unmarried woman as a species of child all her life. Her younger contemporary, Florence Nightingale, later expressed the anguish of that situation: 'Why have women passion, intellect, moral activity — these three — and a place in society where no one of the three can be exercised?' She then observed with gloomy relish: 'Marriage is the only chance (and it is but a chance) offered to women for escape from this death; and how eagerly and how ignorantly it is grasped.' For Caroline the chance had evaporated into thin air. In the Hampshire days she had cared deeply for her cousin Charles, who was of a much more reliable and stable temperament than his elder brother; there had, in fact, been an understanding between the cousins. They shared a love of the countryside and an amused appreciation of their families' foibles. But the debts and difficulties which led to the loss of Hale cast a shadow over the relationship and insensibly drove a wedge of reserve between them. They saw one another less and less frequently, and could talk less and less freely each time they met. When Charles withdrew with the rest of his family to Versailles, Caroline felt relief as well as pain. He never married; but what was she to do with the years that stretched ahead? She was as aware as Florence Nightingale of the danger of 'going mad for want of something to do' that threatened her future.

Fourteen years were spent at Ruxbury, the first fourteen years of Queen Victoria's reign; the political situation, with its Free Trade controversy, Irish Famines and Chartist Movement, seemed less stable at the time than it has seemed in retrospect. Certainly these years were in many ways difficult ones for the widow and her four daughters. Their financial situation compelled them to live 'quietly', compared to the stimulating social life they had led in the orbit of Hale. A list of subscriptions towards the building of a new church at Lyne seems to indicate that the family's income was much reduced from what it had been at Breamore, and compared unfavourably with their neighbours. Mrs May and her daughters could only contribute £8 towards the worthy cause, while Lady Hotham was able to give £50, and Mr Briscoe of Fox Hills £500. By the general standards of their period and social background, three female servants were probably the decent minimum with which such a household could be run, and considered as indispensable as electricity in a modern home; it must be remembered that all domestic tasks were infinitely more strenuous and labour-intensive. Elaborate entertaining was out of the question for the Mays, though occasionally they entertained a cousin or some other visitor. One such in 1841 was Frances Agnew. She was the daughter of a Major Agnew who had been left for dead on the battlefield of Vittoria in the Peninsular War; luckily his Highland servant went out the next day and found the Major under a 'heap of slain'. Even the three successive amputations of his leg without anaesthetic — surgical conditions were primitive — were insufficient to finish him off, and he returned to England where he married a very pretty girl. She always maintained that she married him 'out of pity', but as they had sixteen children (of whom Frances was the second) the emotion must have been singularly enduring.

The Major caused some consternation when he and his wife first called on Mrs May; after the initial civilities he observed abruptly: 'You must be wondering about my leg, Ma'am.' Mrs May's startled and involved reply, in which she strove to combine civil interest with an emphatic disclaimer as to ever having wondered about any gentleman's legs in her life, was long treasured by her family. Frances, however, was less unsettling to entertain than her father and became a frequent visitor; sometimes there were as many as six unmarried young ladies about the house, wearing the thick skirts (heavy with braid and elaborate with frills and tucking) which had become fashionable since 1830; all trying to find occupations that justified their existence.

Here Caroline was fortunate; she had a resource which enabled her to face her bleak single future, her mother's occasional querulousness, her sisters' bickerings and the family's carefully camouflaged economies, with equanimity. For, even before her father's death, Caroline had discovered that she possessed a unique talent for painting flowers. She had been painting since her youth in the Breamore nursery, of course, but the first paintings she kept dated from visits to Dawlish and Teignmouth in Devon in June 1831. After that she painted every flower she could find, but always with a dedicated attention to detail — right down to hairs on a stem and veins in a petal — with a scrupulous eye on the way the plant naturally grew (however far that transgressed the conventional canons of nineteenth-century taste), and a consummate ability to make a beautiful composition out of the most undistinguished little plant. By the time she was at Ruxbury, and cousin Charles was given up for good, a project had formed itself in her mind — nothing less than painting a complete flora of England. With this in mind she went to London to obtain the best Winsor and Newton paints and the very best quality Whatman paper. Her sisters might raise their brows at her extravagance, but Caroline felt this step symbolized a new sense of direction in her life.

So, for more than forty years she painted; sometimes only twenty paintings a year, sometimes many more. What did it matter that nobody took her project seriously (the critic John Ruskin would hold to his opinion that 'no woman could paint' until 1875, the year after Caroline died). For her it was the centre and fulfilment of her life, and in the 1840s her life would have been very drab without it.

But 1850 saw yet another change in fortunes. Henry had long been talked of by his sisters as established for life as an Oxford don. For nearly thirteen years he had applied himself to Classics with devotion, but now a quite different influence disturbed his steady routine. Over the eleven years that he had known Frances Agnew as a friend of the family his feelings for her had steadily deepened. He was 36 and she was 32, but he had made up his mind and within a few months he had resigned his Fellowship at New College in order to marry her. It was not until much later that Oxford Fellows were allowed both wife and a Fellow's seat at High Table.

Edward Archer, a close friend who had been at Winchester and Oxford with Henry, was able to use his influence to get Henry the living of South Petherwyn in Cornwall, not far from the Archer family home at Trelaske, near Launceston. Here Henry and Frances moved on their marriage in 1850, and here, a year later, they proposed that his widowed mother and four unmarried sisters should join them. Mrs May, though in her 71st year, felt all the benefits of such a change in their situation, and Henry found himself at the head of an extensive household. Life in Cornwall suited them all; they made friends, and Mrs Archer's diary records that the Mays were regular guests at Trelaske.

The feminine bias of the May family was to shift as the years went by. Caroline's sister Elizabeth married the Rev. John Sandford in 1852 and moved away from Cornwall; Mrs May herself died in 1855; and Frances presented Henry with six sons: Thomas (born in 1851), Henry (1853), Arthur (1854), Charles (1855), Frederick (1857), and Edmund (1860). The South Petherwyn stipend had to go a long way. As the nursery increased the unmarried aunts decided that they too must help in whatever way they could; they set themselves to study Latin and Mathematics, so that they in turn could undertake the education of their nephews. This they did to such good effect that the boys obtained scholarships to Winchester and Sherborne, and acquitted themselves well in later life: Thomas helped to found Kelly College in Devon and married Mary Gurney,

South Petherwyn vicarage, Cornwall.

a member of the banking family whose wealth was proverbial (as Gilbert and Sullivan admirers will recall); Henry became the youngest Rear-Admiral in the Navy; Arthur was involved in the planning of naval hospitals at Chatham, Devonport and Dartmouth, and eventually became Surgeon-Admiral at the Admiralty and one of King George V's doctors; Charles became head of a firm of solicitors in Lincoln's Inn Fields; Fred followed his father and grandfather as a country clergyman; Edmund became a horticulturalist at Kew.

But all this was well in the future, and meanwhile Caroline painted. Young Katie Archer, daughter of Edward, visiting the household in the 1860s, could remember the three sisters sitting round a table, Louisa and Anna Maria doing exquisite embroidery, and Caroline painting. In accord with the frugal standards they set themselves, the three worked round a single candle. The flower that Caroline was painting was in a jar of water before her, and it was the same jar of water that she was using to mix her paints.

In spite of the rather oppressive mid-Victorian rectitude, it was a happy household and a lively one. Whereas in the Chertsey period it had seemed only too probable that the family would die out within a generation, the six May boys seemed a real guarantee that the future was now assured. In one respect only there was an almost obsessive reticence: Hale was never mentioned, and the boys grew up knowing nothing of the family's background. But, in any case, there was little time now to think of the past. In spite of the economy that reigned at the vicarage it was necessary to engage in a programme of charitable work throughout the parish. Henry's scheme for improving the school in the village became a reality, and his family united in doing all they could to broaden the education of the village children. It also became an established custom for the widows of the parish to call at the vicarage with a pudding-basin on a Sunday morning and take away a 'hot dinner'. The Mays approved Samuel Smiles' advice in *Self-help* (published in 1859) that alms should not be dispersed to the poor in the form of money — that commodity was scarce enough in the vicarage in any case. Then there was the garden with its shrubberies, and the poultry-yard, to be seen to. For the nephews, life in the country provided many opportunities for hobbies and sport. Henry was not addicted to hunting as 'Handsome Thomas' had been (his 'hunting pink' was often only regretfully laid aside at the vestry before a Service), but his sons were allowed to fish and shoot to their hearts' content, and a range of unconventional pets enlivened the household. Of these, a raven learnt to summon their driver, John Blatchford, with such an authoritative imitation of Henry as to cause the dog-cart to be brought to the door unnecessarily on several occasions. On the whole John found it a relief to drive Miss May out, though he regarded her absorption in 'weeds' as eccentricity carried to the point of madness, and was always deaf to her first request that he should rein in Bob, the steady little cob, so that she could get down and examine a possible specimen.

A family photograph, taken in 1870, shows the

whole family grouped outside the South Petherwyn Rectory. Henry and Frances are placed on the left, Caroline sits regally on the right with Louisa and Anna Maria standing on either side of her. The six May boys are placed carefully, each in his way trying to maintain the stillness necessary for the interminable exposure. At first glance Caroline looks forbidding in the extreme, having donned a widow's cap (to which she was not strictly entitled) for extra majesty. But the rigid stiffness of attitude is belied by a humorous twinkle that seems to indicate her appreciation of the figure she makes. Perhaps she didn't suffer fools gladly, but her sixty-one years of varied experience had taught her to season her judgements with compassion and amusement.

By 1871 something almost incredible was happening: Caroline was running out of flowers to paint. In forty years she had painted every wild flower in the areas in which she lived; she had been four times to the Channel Islands and painted the flora there, making the crossing by paddle-steamer from Bridport; she had painted grasses, mosses and lichens; friends had sent her specimens by post from elsewhere in the British Isles. Though now in her sixties her eyesight remained good and her hand steady. Only those paintings executed a few months before her death on 11th December, 1874, show any dimming of her vision or declining of her skills. She is buried in the family grave by the church at South Petherwyn.

She left five great volumes, each of whose brown pages are provided with hand-cut corner slits in the manner of early photograph albums, so that the paintings could be removed or rearranged. In total there are nearly a thousand watercolours (of which the ones in this book represent less than a tenth). The albums were left to her nephew Fred — he is proudly wearing his first bowler hat in the family photo — whom she considered the best botanist of the family. Twenty-three years after her death he married Katie Archer of Trelaske, and the 'Flower Books' were passed down through their daughter, Grace May, to the present owners. Although greatly valued they were not at first treated with especial reverence, but used for general botanical reference by different members of the family. But now, one hundred and sixty years after Caroline first formed the great project that was to be her life's work, a selection of the paintings can be fully appreciated by the public for the first time.

John Tyler

In the only known photograph of Caroline May, she sits between her two unmarried sisters, in this family portrait taken outside South Petherwyn vicarage in 1870.

List of Plates

Captions to the paintings are printed on the page facing the illustration. They start with the modern English and scientific names, with nomenclature according to the standard work *English Names of Wild Flowers* by J.G. Dony, C.M. Rob & F.H. Perring, 1974. Then follows a transcription in italic type of Caroline May's own caption (which often includes scientific names which are now obsolete), and the Linnaen family grouping, sometimes followed by a number. Then the number of the illustration and the volume in which it is found, according to her final order.

Plate 1

Traveller's joy ***Clematis vitalba***

Clematis Vitalba, Travellers Joy, Virgins Bower
Hedges, Breamore, August 1834
Polyandria Polygynia

No. 1, Vol 1

Hedges Braemore

Augst 1834

Clematis Vitalba

Travellers Joy - Virgins Bower

Polyandria Polygynia

Plate 2

Greater celandine ***Chelidonium majus***

Chelidonium Majus, Greater or common Celandine
Shrubbery, Breamore, June 1835
Polyandria Monogynia 112

No. 33, Vol 1

Chelidonium Majus
Greater or common Celandine
Polyandria Monogynia
112
Shrubbery
June 1835

Plate 3

Yellow horned-poppy ***Glaucium flavum***

Chelidonium Luteum, Yellow Horned Poppy
Sea Beach, Ryde, Isle of Wight, August 27, 1836
Polyandria Monogynia 173

No. 34, Vol 1

Chelidonium Luteum.
Yellow Horned Poppy.
Polyandria Monogynia.
Sea Beach Ryde.
Isle of Wight
Augst 27th 1836
173

Plate 4

Cuckooflower ***Cardamine pratensis***

Cardamine Pratensis, Meadow Ladies' Smock
Meadows, Breamore, May 1834
Double Variety in The Orchard at the Vicarage,
South Petherwyn, Cornwall, May 1852
Tetradynamia Siliquosa 39

No. 52, Vol 1

The double variety, added eighteen years after the original painting, is rare in the wild, but less so in the West Country. It was also a favourite cottage flower in Victorian times, and Caroline's orchard specimen may possibly have been an escape from the vicarage flower borders.

Meadows Bucmore Cardamine Pratensis
May 1834
Meadow Ladie's Smock
Tetradynamia Siliquosa

Double Variety
in the Orchard at
The Vicarage South Petherwyn
Cornwall
May 1852

Plate 5

Garlic mustard ***Alliaria petiolata***

Ersymum Alliaria, Jack-by-the-Hedge, Sauce-alone,
Garlic Hedge Mustard
Hedges, Breamore, June 5, 1837
Tetradynamia Siliquosa

No. 59, Vol 1

Ersymum Alliaria
Jack by the Hedge. Sauce alone. Garlic Hedge Mustard
Tetradynamia Siliquosa
Hedges.
Breamore
June 5th 1837.

Plate 6

Wild radish ***Raphanus raphanistrum***

Raphanus Raphanistrum, White-flowered or
Jointed Charlock
Corn fields, Chertsey, Surrey, June 1st, 1841
Tetradynamia Siliquosa

No. 87, Vol 1

Of the two colour varieties figured here, the white-flowered is more common. Wild radish is one of the few arable weeds to have maintained its populations in recent years. To paint the white petals Caroline used no white paint, and creates the illusion of their being whiter than the background by the grey veining.

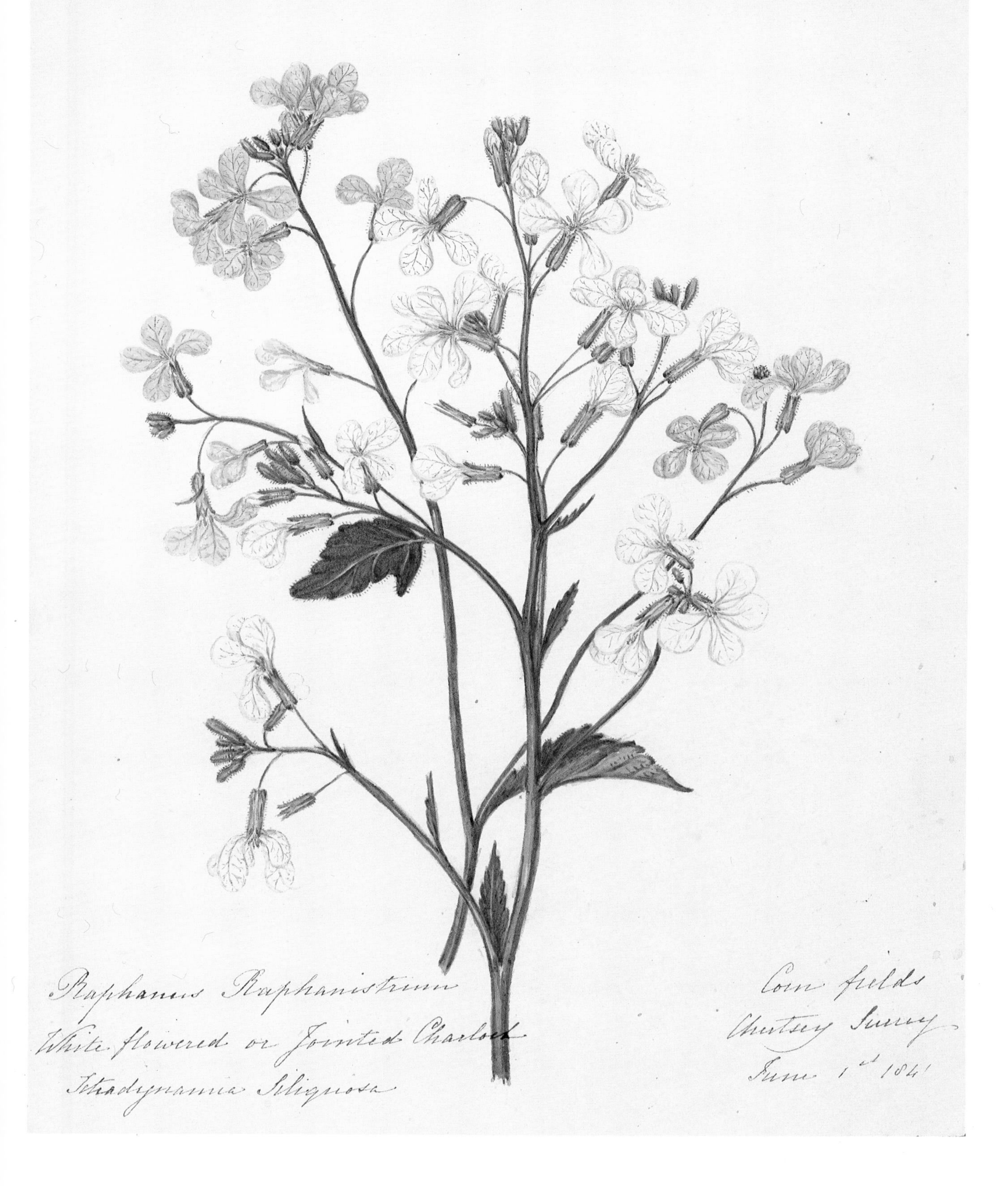
Raphanus Raphanistrum
White flowered or Jointed Charlock
Tetradynamia Siliquosa
Corn fields
Chertsey Surrey
June 1st 1841

Plate 7

Common dog-violet ***Viola riviniana***

Viola Canina, Dog's Violet
Hedges, Breamore, April 1835
Pentandria Monogynia 107

No. 95, Vol 1

Hedges. Brunne — Viola Canina
April 1835 — Dog's Violet

Pentandria Monogynia

Plate 8

Field pansy (smaller flowers) ***Viola arvensis***
and **Wild pansy** ***Viola tricolor***

Viola Tricolor, Wild Pansies, Hearts-ease,
Love in Idleness
Fox Hills and Silverlands, Surrey, July 1842,
Fields, Breamore, May 1834
Pentandria Monogynia 42

No. 98, Vol 1

Caroline May did not distinguish between the two species in her caption. So well integrated are they in this exquisite composition that it is hard to believe that the painting was completed with the addition of the larger flowers and their leaves over eight years after it was begun. The wild pansies, *Viola tricolor*, at the top of the illustration, show the range of the colour varieties that can be found in this species of grassland and (though rarely now) arable fields. Cultivated pansies had begun to be bred from wild stock in 1817, but were still comparatively rare and expensive plants in 1834.

Fox Hills, & Silverlands
Surrey. July 1842
Fields Bremore
May 1834

Viola Tricolor
Wild Pansies. Hearts-ease. Love in Idleness
Pentandria Monogynia
42

Plate 9

White campion ***Silene alba***

Lychnis Dioica, White-flowered Wild Campion
Fields, Breamore, July 1834
Decandria Pentagynia 124

No. 111, Vol 1

The pink-flowered variety is probably the hybrid that occurs between the white campion and the red, *Silene dioica.* The green mark on the right border was almost certainly made when the original was removed from the album for copying by a relative. It would have been quite unlike the artist to have made or left this flaw.

Lychnis Dioica
White flowered Wild Campion
Decandria Pentagynia
124
Fields Braemore
July 1834

Breamore & Hale

Caroline May was a girl of 17 when Greenwood's map, shown here to a scale of 1 inch to the mile, was published in 1826. It shows Breamore three miles north of Fordingbridge in the northern tip of Hampshire — adjacent parts of Wiltshire are left blank.

At Breamore Parsonage, ringed on the map, Caroline was born and lived until she was 28. The house remains today much as it was when sketched, above, in 1842, and coloured pink on the contemporary plan, below. From here Caroline would go to her father's beautiful churches at Breamore, a mile down the lane, and over the river Avon at Hale. From her garden she could see Hale Park across the valley, where her uncle and cousins lived. The May family was an archetype of late eighteenth-century gentry: the head of the family lived in the great house, the second son went into the church and held the local living. Readers of Jane Austen, a contemporary who lived in the same county, will be able to imagine clearly the life of the Mays.

Extending for forty acres away from the parsonage is Breamore Marsh, remarkably unchanged from Caroline's time. Many of the wild flowers Caroline sought here and at nearby Breamore Wood and Breamore Down can surely still be found, for the developments at Woodgreen and Fordingbridge have left the source of Caroline's introduction to botany untouched. She would have crossed the Avon valley by the bridge near the mill, combing the water meadows for plants, and probably encouraged by Gilbert White's *Natural History of Selborne* published first in 1788. Selborne lies just 30 miles to the east.

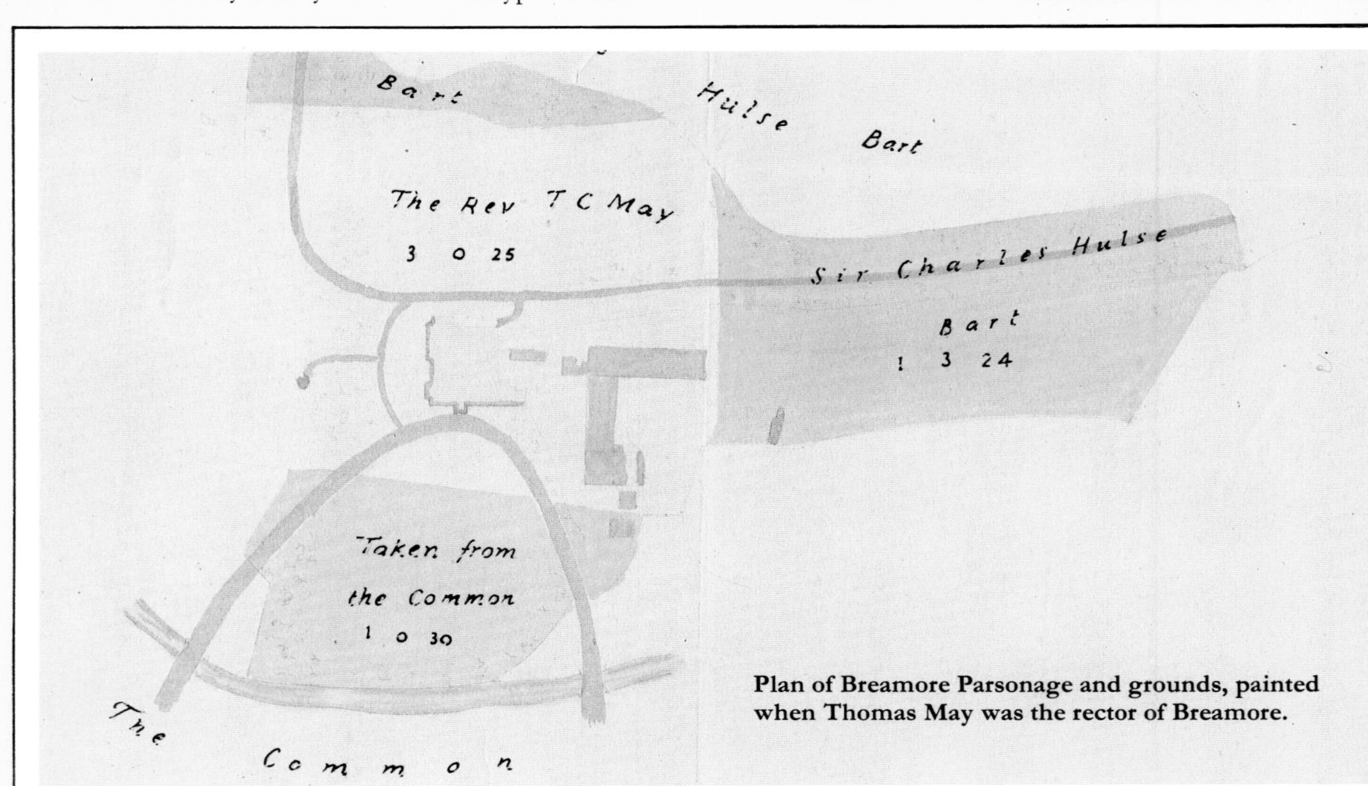

Plan of Breamore Parsonage and grounds, painted when Thomas May was the rector of Breamore.

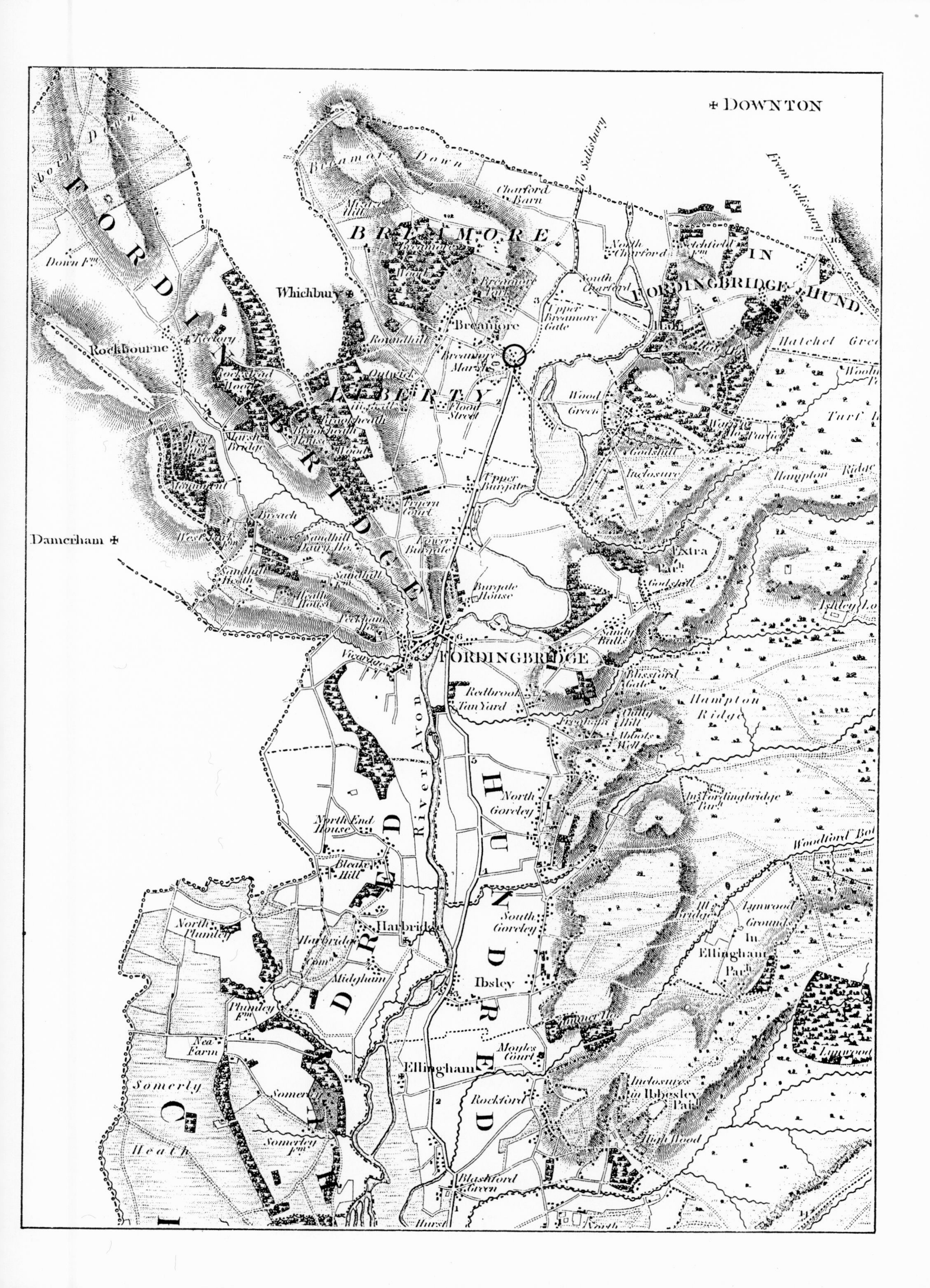

DOWNTON
To Salisbury
From Salisbury
Breamore Down
Charford Barn
BREAMORE
North Charford
South Charford
FORDINGBRIDGE HUND.
Whichbury
Breamore
Upper Breamore Gate
Rockbourne
Roundhill
Outwick
Breamore Marsh
Wood Green
Hatchet Green
LIBERTY
Flood Street
Godshill
Inclosure
Hampton Ridge
Upper Burgate
Damerham
Lower Burgate
Burgate House
Extra Par.
Godshill
Sandy Balls
FORDINGBRIDGE
Vicarage
Redbrook Tan Yard
Hampton Ridge
Abbots Well
River Avon
North Goreley
North End House
Bleak Hill
Harbridge
South Goreley
North Plumley
Midgham
Ibsley
Plumley Fm.
Nea Farm
Moyles Court
Ellingham
Somerly
Heath
Rockford
Inclosures in Ibbesley Par.
High Wood
Somerley Fm.
Blashford Green
Hurst
North
Ellingham Par.
Ill Bridge
Lynwood Grounds
Woodford Bot.
Ashley Lo.

Plate 10

Corncockle ***Agrostemma githago***

Agrostemma Githago, Corn Cockle
Corn Fields, Breamore, July 1834
Decandria Pentagynia 84

No. 114, Vol 1

Now an exceptionally rare cornfield weed throughout Britain, eliminated by modern herbicides and seed-cleaning techniques.

Agrostemma Githago
Corn Cockle
Decandria Pentagynia
84

Corn Fields Bremen
July 1834

Plate 11

Ragged-Robin ***Lychnis flos-cuculi***

Lychnis Flos-cuculi, Ragged Robin, Meadow Pinks,
Wild Williams
Meadows, Breamore, June 1834
Decandria Pentagynia 46

No. 115, Vol 1

The pale variety is quite scarce. Meadow Pinks is a name that Caroline may have learned in the West Country; Wild Williams was the commoner alternative in southern England, and was known to the writer John Gerard in the late sixteenth century.

Lychnis Flos cuculi

Meadows. Bremen

June 1834

Ragged Robin Meadow Pinks Wild Williams

Decandria Pentagynia

46

Plate 12

Imperforate St John's-wort
Hypericum maculatum

Hypericum Dubium, Tutsan, Imperforate St John's Wort
Thorpe Fields, Surrey, July 18, 1840
Polyadelphia Polyandria

No. 147, Vol 1

Caroline May appears to have had some doubt about the identity of this species. The illustration most closely resembles the species for which she gives both an English and a now obsolete scientific name — imperforate St John's-wort (*H. maculatum*, previously *H. dubium*). The short sepals and projecting flower-buds really rule out its close cousin tutsan, *H. androsaemum*. I can find no evidence of the two species sharing a common popular name in any of the areas known to Caroline May.

Hypericum
Polyadelphia Polyandria
Thorpe Fields
Surrey
July 18.. 1840.

Plate 13

Dwarf mallow ***Malva neglecta***

Malva Rotundifolia, Mawls, Dwarf Mallow
Road Side, Mill, Breamore, September 1835
Monadelphia Polyandria 87

No. 159, Vol 1

Malva. Rotundifolia
Mauls. Dwarf Mallow
Monadelphia Polyandria

Road Side. Mill
Breemore
Sep: 1835

87

Plate 14

Dusky cranesbill ***Geranium phaeum***

Geranium Phaeum, Dusky Cranes-bill
Cartha Martha, Cornwall, June 1st, 1857
Monadelphia Decandria

No. 167, Vol 1

Geranium Phaeum
Dusky Cranes-bill
Monadelphia Decandria
Cartha Martha
Cornwall
June 1st 1857.

Plate 15

Orange balsam ***Impatiens capensis***

Impatiens Fulva, Yellow Balsam,
Quick in Hand, a naturalized American plant
Sides of the Canal, Weybridge, Surrey,
September 15, 1843
Pentandria Monogynia

No. 186, Vol 1

This is a fascinating record, made only a few miles from where the species was first seen growing wild in Britain in 1822. The site then was the River Tillingbourne (a tributary of the Wey), and the finder John Stuart Mill, then just sixteen, but shortly to become one of Europe's most influential philosophers. Orange balsam is a North American native, and was first introduced in Britain as a garden plant. But its escape into Surrey's rivers and canals marked the beginning of a rapid colonization of the waterways, made possible by the plant's highly efficient seed dispersal mechanisms. The ripe pods of all members of the balsam family have the habit of bursting violently when touched, and hurling their seeds up to 6 feet (2 metres) away. Hence the Latin family name of *Impatiens*. Orange balsam's seeds have the extra advantage of being light and corky enough to float on water. Launched into the water the seeds float off like tiny coracles, until they lodge in a bank.

'Quick in Hand' was the local West Country name for another balsam, Touch-me-not, *I. noli-tangere*, and was presumably borrowed for this new arrival until its North American name 'Jewelweed' became better known. It is interesting that Caroline May's original caption is just decipherable as *Impatiens Noli-tangere*, and that the *Noli-tangere* was partly erased and replaced with *Fulva*. The fact that at the same date as this correction she added 'a naturalized American plant' indicates that she was aware of her original error.

Impatiens Fulva
Yellow Balsam. Quick in Hand
Pentandria Monogynia
a naturalized American Plant
Sides of the Canal
Weybridge Surry
Sep 15 1843.

Plate 16

Spindle ***Euonymus europaeus***

Euonymus Europaeus, Spindle Tree, Prick-wood
Breamore, 1834, the flower from Petherwyn, Cornwall,
where the flowers have four petals only, 1852
Pentandria Monogynia

No. 192, Vol 1

Caroline May's note about the number of petals is odd, as spindle flowers almost invariably have only four. Prickwood is a vernacular naming which derives from spindlewood's use in making skewers. Caroline clearly knew that she would add a painting of the flowers to the study of the seeds done in 1834, which is why she placed these seeds so far to the left.

Euonymus Europæus
Spindle Tree. Prick-wood
Pentandria Monogynia.
Bremore
1834

Plate 17

Alder buckthorn ***Frangula alnus***

Berries of Rhamnus Frangula, Alder Buckthorn
Pentandria Monogynia

No. 196, Vol 1

Berries of
Rhamnus Frangula
Alder Buckthorn
Pentandria Monogynia

Plate 18

Gorse ***Ulex europaeus***

Ulex Europaeus, French Furze, Gorse or Common Furze
Woodgreen Heights, New Forest, April 1833
Diadelphia Decandria 98

No. 197, Vol 1

Woodgreen Heights
New Forest
April 1833

Ulex Europeaus
French Furze Gorse or Common Furze
Diadelphia Decandria
38

Plate 19

White clover ***Trifolium repens***

Trifolium Repens, Variety of Dutch Trefoil
Lane between Staines and Chertsey, August 3, 1845
Diadelphia Decandria

No. 221, Vol 2

Caroline's pencil caption has been corrected but never inked in. No white paint was used for the flowers.

Trifolium Repens
Variety of Dutch Trefoil
Diadelphia Decandria

Lane between
Staines & Chertsey
Aug 3 18[illegible]

Plate 20

Tufted vetch ***Vicia cracca***

Vicia Cracca, Tufted Vetch
Lanes at Breamore, August 1835
Diadelphia Decandria 156

No. 239, Vol 2

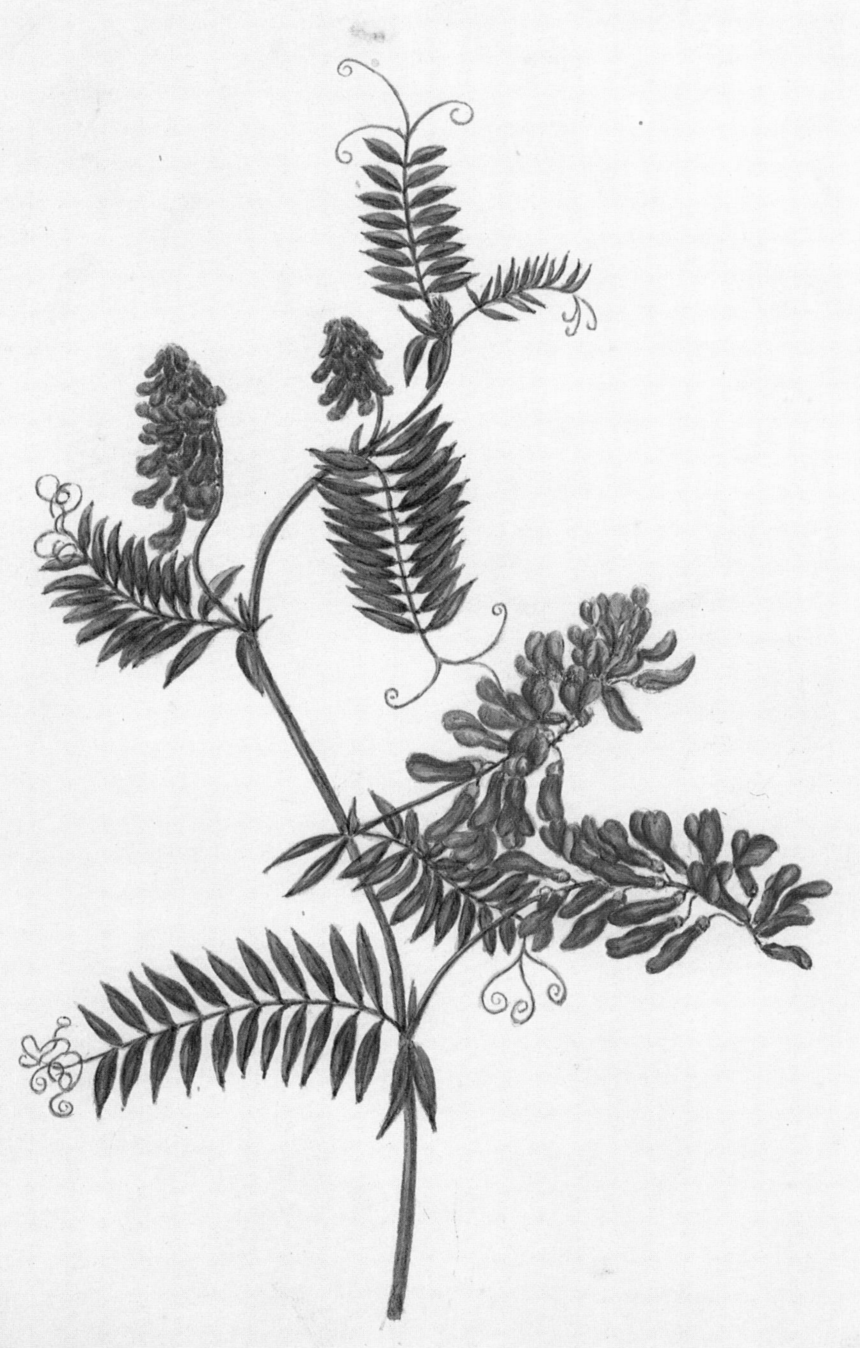

Vicia Cracca
Tufted Vetch
Diadelphia Decandria

Lanes at Braemar
Augt 1835
156

Plate 21

Wood vetch ***Vicia sylvatica***

Vicia Sylvatica, Wood Vetch
Copse near Market Lavington, Wilts, July 1864
Diadelphia Decandria

No. 240, Vol 2

Woods on the Wiltshire chalk are one of the strongholds of this scarce species.

6

Vicia Sylvatica
Wood Vetch
Diadelphia Decandria

Copse near
Market Lavington
Wilts. July
1864

Plate 22

Narrow-leaved everlasting-pea ***Lathyrus sylvestris***

Two varieties of Lathyrus Sylvestris
Narrow-leaved Everlasting Pea
Red Variety, Hedge skirting Breamore Wood, August 3, 1837
Purple, Wood at Clyffe Hall, Wilts, August 26, 1845
Diadelphia Decandria

No. 248, Vol 2

The flowers of this rather scarce southern species do vary in colour, and also fade during blooming.

two varieties of
Lathyrus Sylvestris
Narrow leaved Everlasting Pea.
Diadelphia Decandria.
Red Variety
Hedge skirting
Breamore Wood.
Augst. 3d. 1837.
Purple — Wood at
Clyffe Hall. Wilts. Aug 26
1846

Plate 23

Blackthorn ***Prunus spinosa***

Prunus Spinosa, Blackthorn, Sloe
Lanes at Breamore, April 1833, the Sloes October 2nd, 1837
Icosandria Monogynia 24

No. 250, Vol 2

Lanes at Bremore
April 1833
the Sloes. Oct 2d 1837

Prunus Spinosa
Blackthorn. Sloe
Icosandria Monogynia

24

Plate 24

Bullace ***Prunus domestica*** subs. ***insititia***

Two varieties of the Fruit of the Prunus Insititia,
Bullace Plumb
Lyne Lane, Surrey, September 22, 1842
Icosandria Monogynia

No. 252, Vol 2

Two varieties of the
Fruit of the Prunus Insititia
Bullace Plumb
Icosandria Monogynia
Lyne Lane
Surrey. Septbr 22.
1842.

Plate 25

Meadowsweet ***Filipendula ulmaria***

Spiraea Ulmaria, Common Meadow Sweet
Banks of the Avon, June 1835
Icosandria Pentagynia 140

No. 256, Vol 2

Meadowsweet was once used medicinally against malaria and other fevers, and was the plant from which salicylic acid (the origin of aspirin) was first isolated in Italy, just three years after Caroline May's painting. Aspirin itself (acetylsalicylic acid) was first formulated in 1899, and was named in honour of meadowsweet's old botanical name *Spiraea*.

Spiraea Ulmaria
Common Meadow Sweet
Icosandria Pentagynia
Banks of the Avon
June 1835

Plate 26

Dewberry ***Rubus caesius***

Rubus Caesius, Dew-berry Bush
Lanes at Breamore, October 30th, 1837
Icosandria Polygynia

No. 266, Vol 2

Rubus Caesius
Dew-berry Bush
Icosandria Polygynia

Lunsat Brunne
Octbr. 30th 1837

Plate 27

Tormentil ***Potentilla erecta***

Tormentilla Officinalis, Sept-foil, Common Tormentil
Silverlands, Surrey, May 28, 1844
Icosandria Polygynia

No. 273, Vol 2

Tormentil is often called cinquefoil because of the groups of five leaflets that surround the stem. 'Sept-foil' was the name for a very rare variety with seven leaflets. The plants drawn here appear to have the usual five leaflets, but Caroline has, as so often, managed to find an exceptional double-flowered variety.

Tormentilla Officinalis
Sept-foil Common Tormentil
Icosandria Polygynia
Silverlands
Surrey
May 28. 1844

Plate 28

Marsh cinquefoil ***Potentilla palustris***

Comarum Palustre, Marsh Cinquefoil
Flowers at Congdon, Sth. Petherwyn, Cornwall, May 1871
Seed Near Baschurch, Shropshire, August 1st, 1857
Icosandria Polygynia

No. 278, Vol 2

Comarum Palustre
Marsh Cinquefoil
Icosandria Polygynia
Flowers at Congdon
St. Petherwyn. Cornwall
seed near Baschurch May 1871
Shropshire.
Aug. 1st 1857

Plate 29

Burnet rose ***Rosa pimpinellifolia***

Rosa Spinosisima, Burnet Rose, Pimpernel Rose
Hedge at Lavington, Wilts, June 2nd, 1835
Icosandria Polygynia 116

No. 287, Vol 2

An unusual record. This species is rare in southern England away from the coast.

Hedge at Lavington
Wilts.
June 2d 1835

Rosa Spinosisima
Burnet Rose. Pimpernel Rose
Icosandria Polygynia
116

Plate 30

Sweet briar ***Rosa rubiginosa***

Rosa Rubignosa, Sweet Briar, Rose Eglantine
Garden, Breamore Rectory, June 1833
Wild at Lyne, Surrey
Icosandria Polygynia 80

No. 288, Vol 2

Garden. Bremore Rectory. June 1833
Wild at Lyne
Surrey

Rosa Rubiginosa

Sweet Briar. Rose Eglantine

Icosandria Polygynia

Plate 31

Crab apple ***Malus sylvestris***

Pyrus Malus, Crab Tree Wilding
Lane between Thorpe and Staines, Middlesex, May 2nd, 1843
Icosandria Pentagynia

No. 296, Vol 2

Pyrus Malus
Crab Tree Wilding
Icosandria Pentagynia
Lane between
Thorpe & Staines
Middlesex
May 2d. 1843

Plate 32

White bryony ***Bryonia dioica***

Bryonia Dioica, Red-berried Bryony, Wild Vine
Church Lane, Breamore, July 22nd, 1836
Triandria Monogynia 164

No. 319, Vol 2

Church Lane
Breamore
July 22. 1836

Bryonia Dioica
Red=berried Bryony. Wild Vine
Triandria Monogynia
164

Plate 33

Hemlock water-dropwort ***Oenanthe crocata***

Oenanthe Crocata, Hemlock Water Dropwort,
Nat. Order Calyciflora Umbelliferae
Ditch, Chertsey, Surrey, June 6, 1848
Pentandria Digynia

No. 354, Vol 2

Probably the most poisonous of all British native plants, and hazardous even to illustrators, if they fail to take proper precautions. When Ehret, the eighteenth-century flower painter, was working with a specimen, 'the smell, or effluvia only, rendered him so giddy that he was several times obliged to quit the room, and walk out in the fresh air to recover himself; but recollecting at last what might probably be the cause of his repeated illness, he opened the door and windows of the room, and the free air then enabled him to finish his work without any more return of his giddiness.' (John Lightfoot, *Flora Scotica*, 1777.) It can be seen under a lens that no white paint was used in the amazing painting of the flowers.

Œnanthe Crocata
Hemlock Water Dropwort
Nat. Order. Calyciflorae Umbelliferae
Pentandria Digynia
ditch. Chertsey
Surrey
June 6. 1848

Ruxbury House

Chertsey

Two motorways now cross the field within a few yards of Ruxbury House, Chertsey, where Caroline May lived from 1838 until 1851. The house, lying on the lower slopes of St Ann's Hill and adjoining the estate where Charles James Fox's widow lived until her death in 1842, is ringed on the reproduction from Froggett's 2 inch to the mile map, published around 1830.

Wholly rebuilt in 1872, Ruxbury House today looks very different from the house with orangery, stables, garden and orchards, which is seen in the drawing above, as it was when Mrs May rented it from a Chertsey landowner. The Mays must already have had friends here when they moved from Hampshire, for Caroline's father's first wife was a daughter of Sir Joseph Mawbey, 1st Bart, who had built himself a handsome house at the neighbouring Botleys Park (below) in about 1765, and Thomas May had been vicar of Chertsey, a Mawbey living, from 1805. Though a small market town, Chertsey was then esteemed for 'the respectability of the neighbourhood' and the gentility of its numerous grander residents. Among the May's circle of acquaintance would surely have been Lady Frances Hotham of Silverlands, where Caroline painted, John Ivatt Briscoe, a Mawbey relative, of Foxhills, and Gilbert White's niece, who lived at Monks Grove.

Many of Caroline's sites are marked on Froggett's map: France Farm, Lyne Lane, Timber Hill, Thorpe, Fox Hills, Ham and Weybridge. Much has inevitably changed in an area now on the fringe of London, but between the railways and motorways there are still leafy lanes and tranquil woods where Caroline walked, looking for flowers, and where one can still do so. Caroline found pennyroyal (see page 6) in Hardwick Lane, on the edge of Silverlands park, but now it is an endangered species, found only in one site in Surrey.

Botleys Park

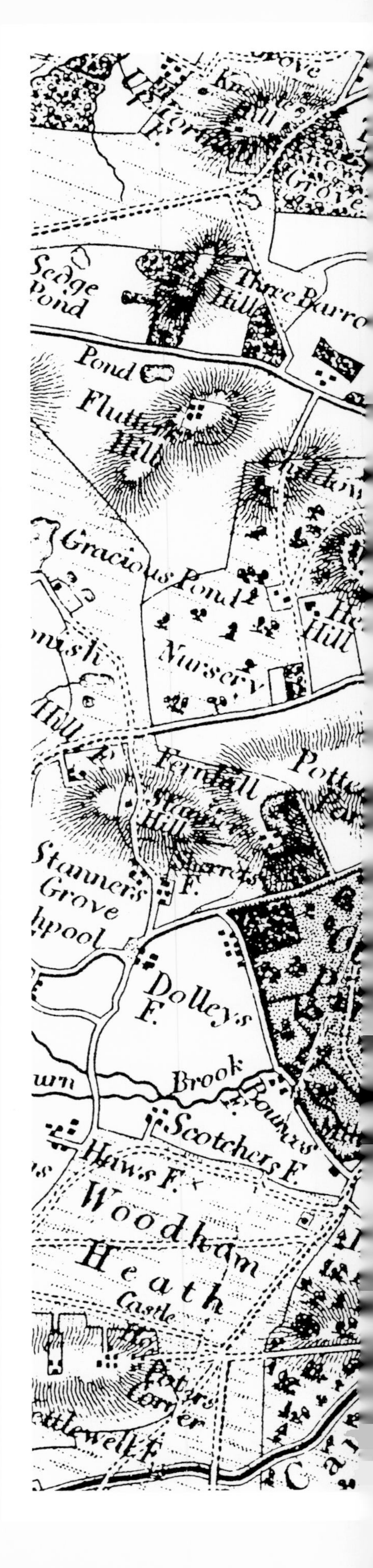

Thorpe
Trumps Mill
Worlds End
Lyne
Mill
Lyne Green
St. Anns Hill
Monks Grove
CHERTSEY
19½
Laleham
Lalehan Park
Ferry
York Ho.
Abbey Mill
Putworth Ho.
Hardwick Court F.
Botleys
Abbey
Beaumont Lo.
Chertsey Br.
18
Shep
Pannels F.
Chertsey Mead
Hatch F.
Simplemarsh F.
Addlestone
Captains F.
The Grove
Spinney Oak
Brooks F.
Bawsley F.
Ongar Hill Cot.
Ham Moor
Iron Mills
Ongar Hill Ho.
Woodham
Weybridge
Wey Riv.
New Haw F.
Brooklands F.
Holly F.
Nursery

Plate 34

Fennel ***Foeniculum vulgare***

Meum Foeniculum, Common Fennel
Hedge Row in a hilly field near Ham, Surrey, August 6, 1848
Pentandria Digynia

No. 357, Vol 2

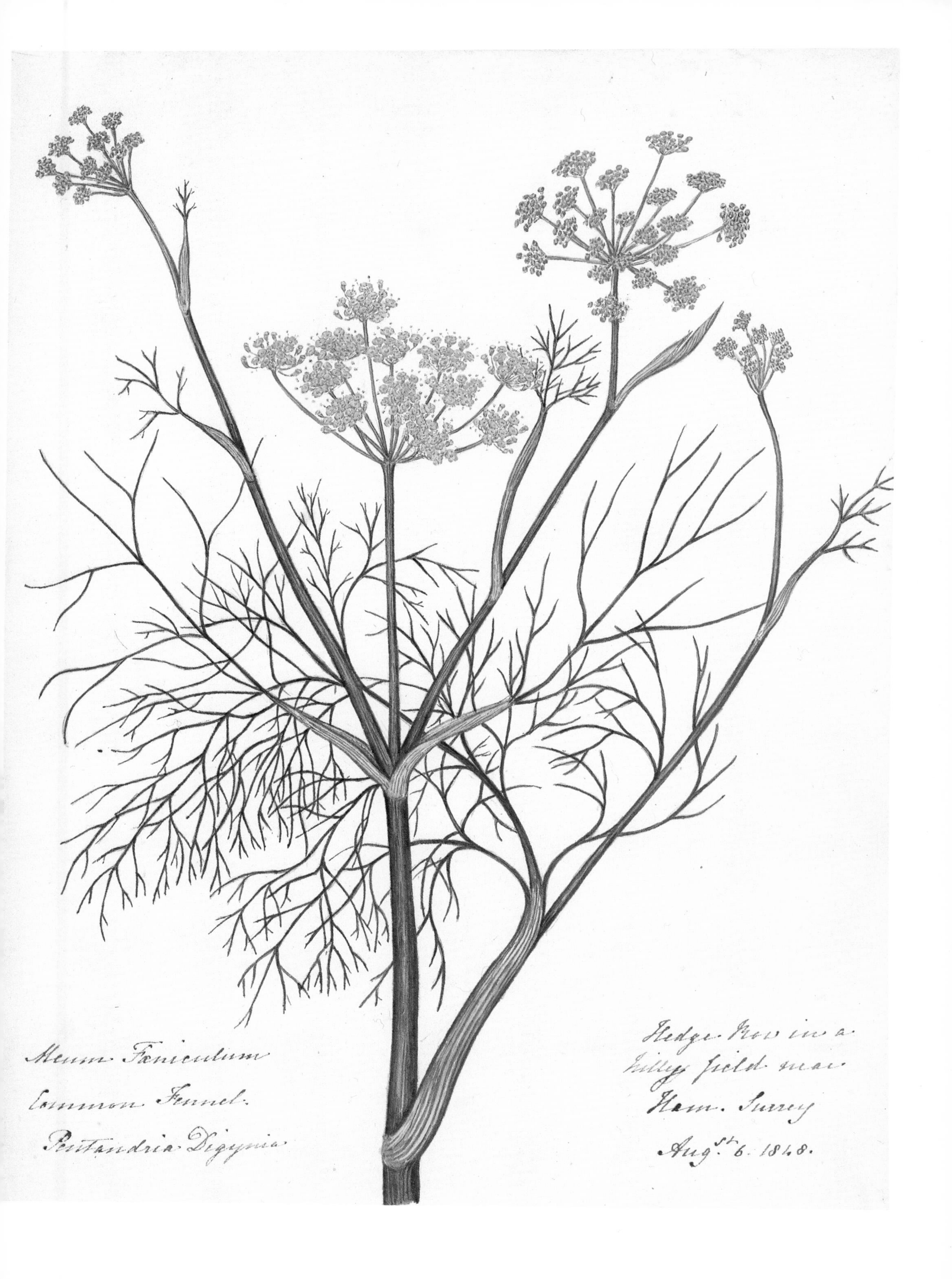

Meum Fœniculum
Common Fennel.
Pentandria Digynia
Hedge Row in a
hilly field near
Ham. Surrey
Aug.st 6. 1848.

Plate 35

Wild angelica ***Angelica sylvestris***

Angelica Sylvestris, Wild Angelica
Wood near the River Inney, Cornwall, August 24, 1857
Pentandria Digynia

No. 360, Vol 2

ngelica
Wild
Pentandri Digynia
Wood the river Inney
Cornwall,
Aug. 24. 1857

Plate 36

Mistletoe ***Viscum album***

Viscum Album, White Misseltoe
Garden at Breamore, April 20th, 1837
Tetrandria Monogynia

No. 377, Vol 2

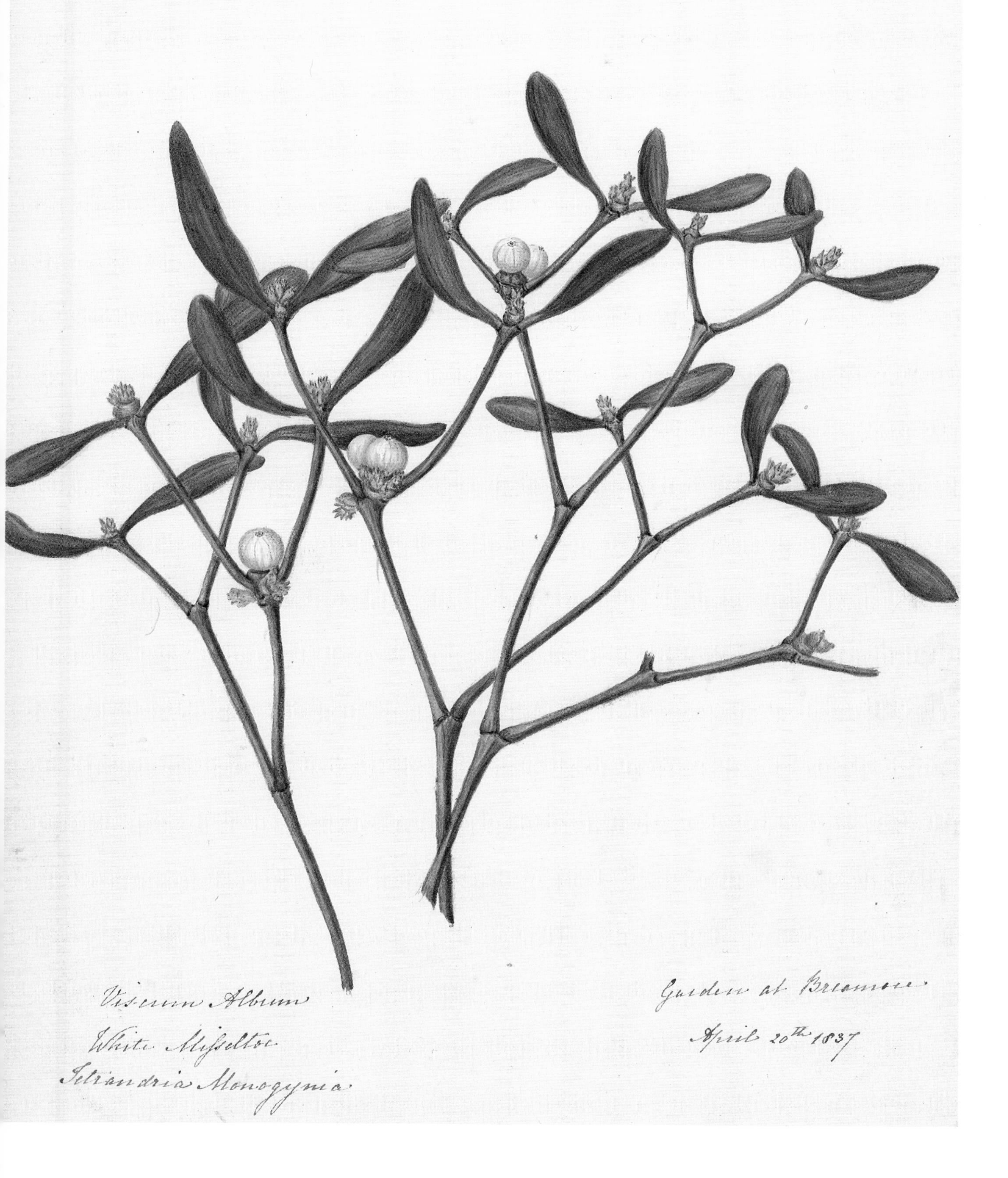
Viscum Album
White Misseltoe
Tetrandria Monogynia
Garden at Breamore
April 20th 1837

Plate 37

Wayfaring-tree ***Viburnum lantana***

Viburnum Lantana, Wayfaring Tree, Mealy Guelder Rose
Lane, Kingston, Surrey, May 29, 1845
Pentandria Trigynia

No. 383, Vol 2

The 'mealy' of the secondary common name refers to the white woolly down beneath the leaves and around the outer shoots. No white paint was used for the flowers, but the black berries were carefully varnished.

Viburnum Lantana
Wayfaring Tree Mealy Guelder Rose
Pentandria Trigynia
Lane. Kingston
Surrey
May 29. 1845

Plate 38

Wild madder ***Rubia peregrina***

Berries of the Rubia Tinctorum, Wild Madder
Woods at Torquay, January 1837.
Flowers (same place) June 1870
Tetrandria Monogynia

No. 387, Vol 2

More or less confined to coastal areas in the south-west, but not uncommon there. The berries had to wait thirty-three years to be joined by the flowers, painted after a much later visit to Torquay. The technique she used for painting the prickles was identical.

Barries of the Rubia Tinctorum
Flowers (same place)
Wild Madder
Tetrandria Monogynia
Woods at Torquay
Jan.y 1837
June 1870

Plate 39

Field scabious ***Knautia arvensis***

Scabiosa Arvensis, Field Scabious
Fields at Breamore, August 1835
Tetrandria Monogynia

No. 413, Vol 3

The picture shows three colour types of a very variable species. Here Caroline used opaque white paint for details of the blue flower, but not for the white one.

Scabiosa Arvensis
Field Scabious
Tetrandria Monogynia
Fields at Braemar
Aug.st 1835

Plate 40

Butterbur ***Petasites hybridus***

Tussilago Petasites, Colts-foot, Butter-bur,
Pestilent-wort
Bank at Sea Lawn, Dawlish, December 1836
Syngenesia Superflua

No. 416, Vol 3

It is highly unlikely that Caroline May confused butterbur and the true coltsfoot *Tussilago farfara* (which has yellow flowers), so her recording of Coltsfoot as an alternative name probably indicates a local naming, derived from the similar leaf-shapes of the two species. Pestilent-wort reflects the ancient use of butterbur root against the plague. In Germany it is still called *Pestilenzwurz*. Opaque white paint was used only for the stigmas.

Tussilago Petasites
Colts-foot. Butter-Bur. Pestilent-wort.
Syngenesia Superflua.
Bank at Sea Lawn
Dawlish.
Decr. 1836

Plate 41

Sea aster ***Aster tripolium***

Aster Tripolium, Sea Star-Wort
Sea Coast, Ryde, Isle of Wight, August 27th, 1836
Syngenesia Superflua 175

No. 418, Vol 3

Aster Tripolium
Sea Star-Wort.
Syngenesia. Superflua
175
Sea Coast. Ryde
Isle of Wight
Augst 27th 1836

Plate 42

Yarrow ***Achillea millefolium***

Achillea Millefolium, Yarrow, Millfoil
Fields at Breamore, July 1834
Syngenesia Superflua 71

No. 439, Vol 3

White is the usual colour of yarrow flowers, and to find three different colour varieties in the same area would have been most unusual. Perhaps, on occasions, Caroline May painted from memory or specimens sent to her from elsewhere, as well as direct experience.

Fields at Breamore
July 1834
Achillea Millefolium
Yarrow. Millfoil
Syngenesia Superflua
71

Plate 43

Saw-wort ***Serratula tinctoria***

Serratula Tinctoria, Common Saw-Wort
Hedge at Warrens, New Forest, August 1st, 1837
Syngenesia Aequalis

No. 459, Vol 3

Now a rather scarce species of heaths and rough grassland. The English name saw-wort refers to the jagged edges of the leaves; the scientific specific *tinctoria* to the strong yellow dye that can be obtained from the leaves.

Serratula Tinctoria
Common Saw-Wort.
Syngenesia Æqualis.
Hedge at Warrens.
New Forest.
Aug.st 1.st 1837.

Plate 44

Creeping thistle ***Cirsium arvense***

Carduus Arvensis, Field Thistle
Road Side, Tipner, Hants, September 23, 1840
Polygamia Aequalis

No. 469, Vol 3

Carduus Arvensis
Field Thistle
Polygamia Æqualis.
Road Side
Tipner Hants
Sep. 23. 1840

Plate 45

Greater knapweed ***Centaurea scabiosa***

Centaurea Scabiosa, Greater Knapweed
Fields near Dawlish, October 24th, 1836
Pink Variety, bank, near Landhu, Cornwall, July 6, 1870
Syngenesia Frustranea

No. 478, Vol 3

Again, the white and pink varieties are very unusual finds.

Centaurea Scabiosa
Greater Knapweed
Syngenesia Frustranea
Fields near Dawlish
Oct. 24th 1836
Pink Variety
bank, near Landhu
July 6. 1870 Cornwall

South Petherwyn

When Caroline May moved to South Petherwyn in Cornwall at the age of 42, she found herself a few miles south of Launceston and barely a parish away from the Devon border. The countryside around the village was a traditional blend of woodland and well-hedged fertile pasture. The church stood at the highest point of the village, with Bodmin Moor to the west and Dartmoor to the east. The vicarage itself was a quarter of a mile down the hill towards Launceston, held in a sheltered hollow of lawns and shrubberies.

South Petherwyn (the Cornish spelling of the name was always variable, but was definitely modified to 'Petherwin' a few years after her death; the map here is dated 1883) has changed relatively little in 140 years, and Caroline would still find her way without difficulty to many of her favourite haunts.

The river Tamar at Cartha Martha (below, by an unknown hand), in the neighbouring parish of Lezant, is still as beautiful as ever, and it was here that she found the dusky cranesbill (Plate 14) and bird cherry (Plate 83). By now her eye was trained by a lifetime of experience, so that she was able to spot elsewhere the very unusual colour varieties of greater knapweed (Plate 45) and germander speedwell (Plate 60).

Over one hundred of her paintings come from Cornwall, showing her enthusiasm unabated until the end of her life. Her figure, absorbed in its searchings among the hedgerows and marsh, moor and woodland, must have been a familiar sight to her brother's parishioners.

She is buried, with her brother Henry, in the graveyard to the south of the church.

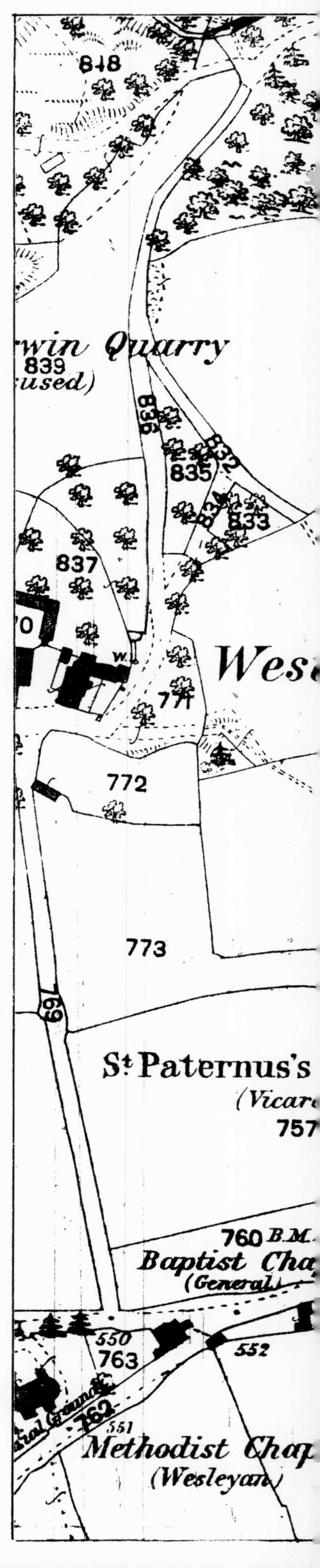

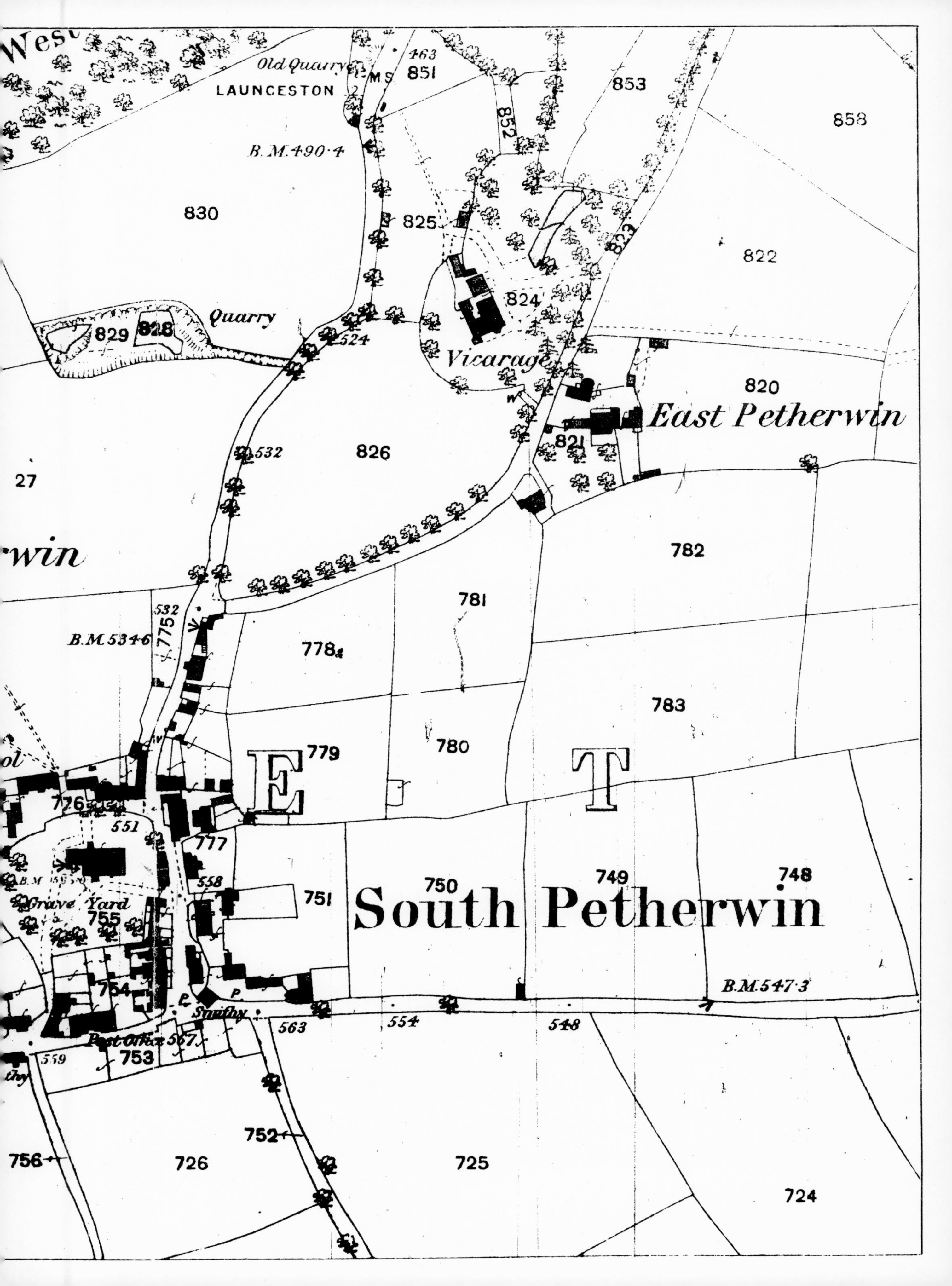

Old Quarry
LAUNCESTON
851
853
858
852
B.M. 490·4
830
825
822
823
824
829
828
Quarry
524
Vicarage
820
East Petherwin
821
532
826
27
782
781
532
775
B.M. 534·6
778a
783
779
780
776
551
777
748
749
750
558
751
South Petherwin
Grave Yard
755
754
Smithy
B.M. 547·3
563
554
548
Post Office 567
753
752
756
726
725
724

Plate 46

Cornflower ***Centaurea cyanus***

Centaurea Cyanus, Blue Bottle, Corn Flower, Knapweed
Cornfields, Breamore, July 1834
Purple variety, Sth. Petherwyn, June 1870
Syngenesia Frustranea 48

No. 479, Vol 3

Now exceptionally rare in Britain.

Corn-fields. Bricamore Centaurea Cyanus
July 1834
Blue Bottle. Corn Flower. Knapweed
Purple variety
Pethroyn
June 1870. Syngenesia Frustranea
48

Plate 47

Clustered bellflower ***Campanula glomerata***

Campanula Glomerata, Little Throatwort,
Canterbury Bells, Clustered Bell Flower
Breamore Wood, September 1834
Pentandria Monogynia 90

No. 508, Vol 3

The name Canterbury bells is usually applied to the giant bellflower, *C. latifolia*, also known as the greater throatwort, of which this is a miniature version.

Campanula Glomerata

Little Throatwort - Canterbury Bells Clustered Bell Flower

Pentandria Monogynia

9°

Bremmer Wood
Sep' 1834

Plate 48

Spreading bellflower ***Campanula patula***

Campanula patula, Spreading or Field Bell Flower
Densham Wood, New Forest, September 1834
Pentandria Monogynia 120

No. 511, Vol 3

A very scarce and declining woodland species.

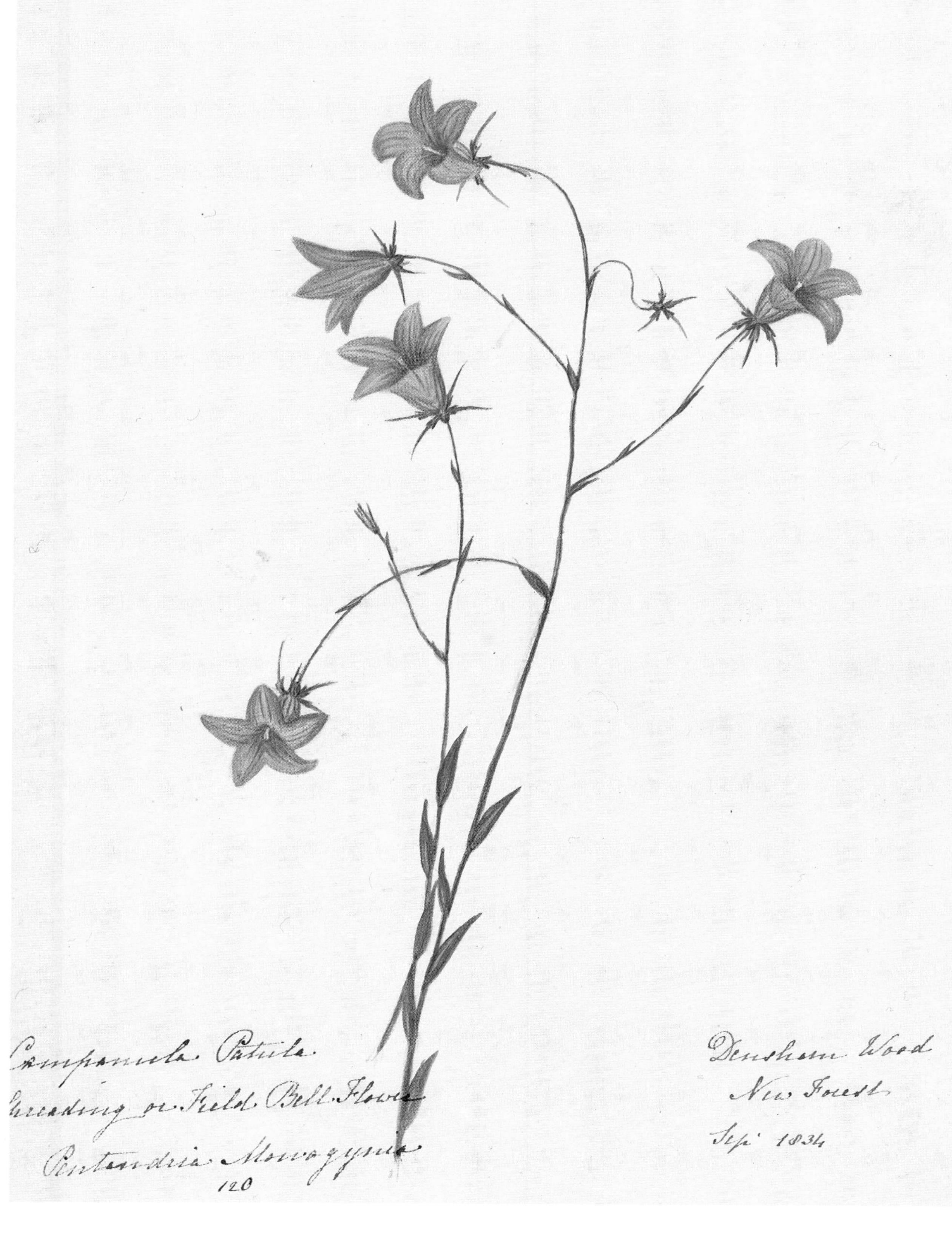
ampanula Patula
preading or Field Bell Flower
Pentandria Monogynia
120
Densham Wood
New Forest
Sep 1834

Plate 49

Cowslip ***Primula veris***

Primula Officinalis, Cowslip
Stanwell, Middlesex, April 1839
Red Variety found in a field at Dawlish, Devon, May 1852
Pentandria Monogynia

No. 526, Vol 3

The red variety, added thirteen years later, has always been associated with the West Country, and is commonly known to gardeners as 'Devon Red'.

Primula Officinalis
Cowslip
Pentandria Monogynia
Stanwell Middle
April 1839
Red Variety found
in a field at
Dawlish - Devon.
May 1852 -

Plate 50

Yellow loosestrife ***Lysimachia vulgaris***

Lysimachia Vulgaris, Yellow Willowherb,
Yellow Loosestrife
Meadows by the River Avon
Pentandria Monogynia 142

No. 529, Vol 3

Lysimachia Vulgaris

Meadows by the river Avon

Yellow Willowherb Yellow Loosestrife

Pentandria Monogynia

142

Plate 51

Greater periwinkle ***Vinca major***

Vinca Major, not really wild, Greater Periwinkle,
Garden at Breamore Rectory
Pentandria Monogynia 22

No. 543, Vol 3

Vinca Major not really wild
Greater Periwinkle
Pentandria Monogynia

Garden at
Bresmore Rectory

Plate 52

Lesser periwinkle ***Vinca minor***

Vinca Minor, Lesser Periwinkle
Garden at Sea Lawn, Dawlish, December 1836
Wild in a hedge at Breamore,
White Variety at Wickham, Hants, March 19, 1844
Pentandria Monogynia

No. 544, Vol 3

Unlike the previous species, which is an introduction to Britain, lesser periwinkle may possibly be a native plant in the south-west of the country. Both species readily naturalize from garden colonies.

Vinca Minor
Lesser Periwinkle
Pentandria Monogynia
Garden at Sea Lawn
Dawlish Dec. 1836.
Wild in a hedge at
Breamore.
White Variety at Wickham
Hants. March 29. 1844

Plate 53

Bogbean ***Menyanthes trifoliata***

Menyanthes Trifoliata, Water Trefoil, Marsh Cleam,
Trefoil Buckbean
Bogs at the Moor, Breamore, June 8, 1837
Pentandria Monogynia

No. 551, Vol 3

'Marsh Cleam' is an unrecorded vernacular name.

Menyanthes Trifoliata

Water Trefoil. Marsh Clover. Trefoil. Buckbean.

Pentandria Monogynia

Bogs at the Moor

Breamore.

June 8. 1837.

Plate 54

Hedge bindweed ***Convolvulus sepium***

Convolvulus Sepium, Great Bind-weed
Breamore Rectory Shrubbery, July 1831
Pentandria Monogynia 14

No. 555, Vol 3

No white paint was used on the flower.

Rectory Shrubbery
July. 1831.
Convolvulus Sepium
Great Bind-weed
Pentandria Monogynia
14

Plate 55

Henbane ***Hyoscyamus niger***

Hyoscyamus Niger, Common Henbane
Road side, Knaphill, Surrey, June 10, 1844
Pentandria Monogynia

No. 576, Vol 3

An occasional denizen of waste places. In Caroline May's time henbane — a member of the nightshade family — was used extensively in medicine. It contains several potent chemicals with narcotic and sedative effects. One, hyoscine, is still frequently prescribed for travel-sickness and intestinal complaints. Opaque white paint was used on some of the flowers and leaves.

Hyoscyamus Niger
Common Henbane
Pentandria Monogynia
Road side
Knaphill
Surrey
June 10. 1844

Plate 56

Black nightshade ***Solanum nigrum***

Solanum Nigrum, Common or Garden Nightshade
Lanes, Breamore, September 1835
Pentandria Monogynia 128

No. 580, Vol 3

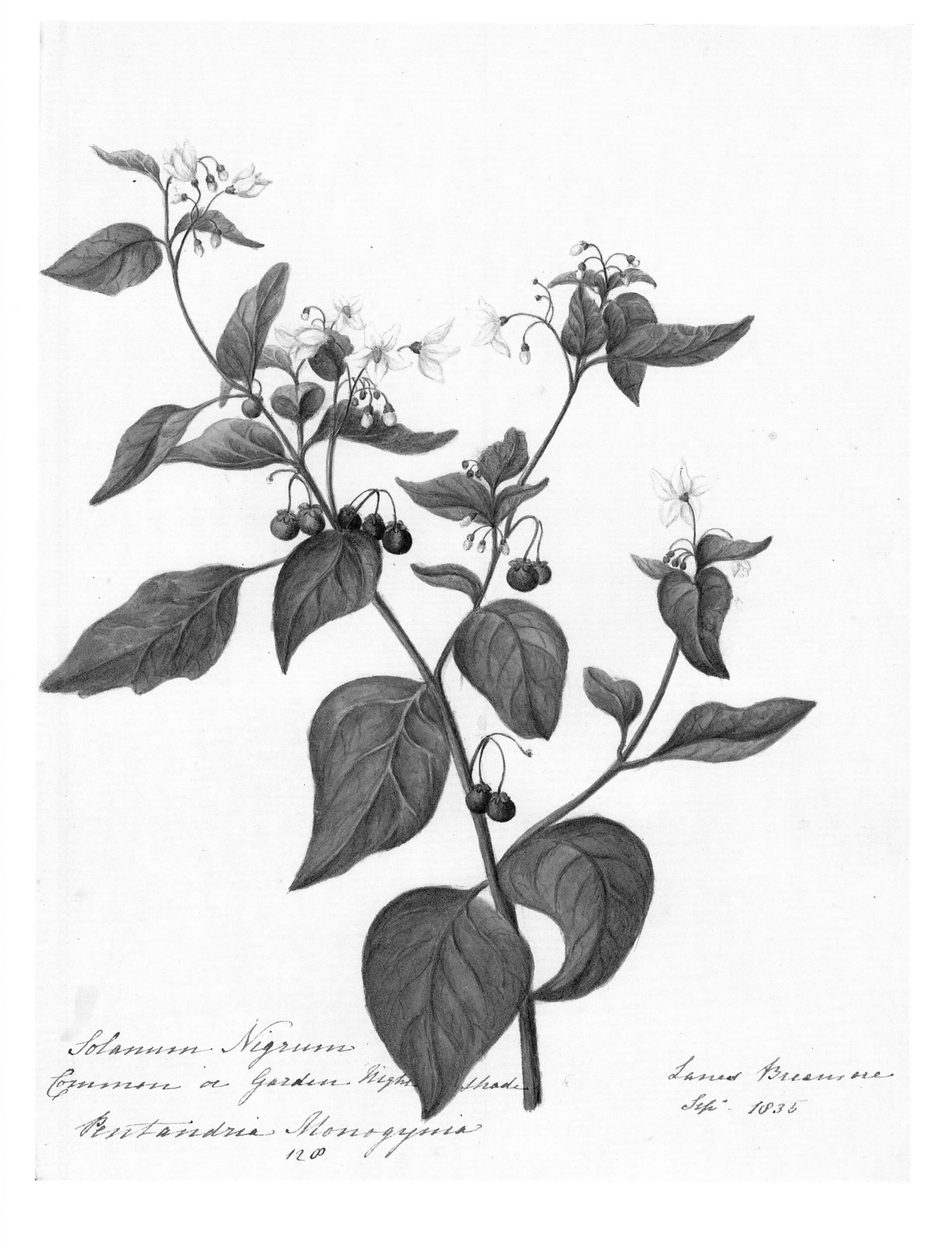
Solanum Nigrum
Common or Garden Night Shade
Pentandria Monogynia
120
Lanes Breamore
Sepr 1835

Plate 57

Deadly nightshade ***Atropa belladonna***

Atropa Belladonna, Deadly Nightshade
Cockshot, Buriton, Hants, July 26, 1844
Pentandria Monogynia

No. 581, Vol 3

Atropa Belladonna
Deadly Nightshade
Pentandria Monogynia

Cockshel. Buriton
Hants
July. 26. 1844.

Plate 58

Twiggy mullein ***Verbascum virgatum***

Verbascum Virgatum, Large-flowered Mullein
Roadside from St Anne's Hill to Chertsey, Surrey,
September 12th, 1838
Pentandria Monogynia

No. 590, Vol 3

Truly wild specimens of this rare species are confined to south-west England, and this specimen discovered by a Surrey roadside was probably a garden escape.

Verbascum Virgatum
Large flowered Mullein
Pentandria Monogynia
Road side from St Annes Hill
to Chertsey Surrey
Sepr 12th 1838

Plate 59

Monkeyflower ***Mimulus guttatus***

Mimulus Guttatus, Monkey Plant,
a naturalized American Plant
Watery Lane near Chobham Common, Surrey,
August 14, 1840
Didynamia Gymnospermia

No. 607, Vol 4

The monkeyflower comes from the Aleutian Islands, off Alaska. The first specimens were in English gardens by 1812, and the first naturalized colonies observed in 1824.

Mimulus Guttatus

Monkey Plant   a naturalized American Plant

Didynamia Gymnospermia.

Watery lane nea

Chobham common

Surrey

Aug. 14.. 1840.

Plate 60

Germander speedwell ***Veronica chamaedrys***

Veronica Chamaedrys, Wild Germander,
Germander Speedwell
Breamore Lanes, May 1832
Light variety S. Petherwyn, May 1856
Diandria Monogynia 27

No. 621, Vol 4

The alternative colour varieties are very scarce.

Breamore Lanes
May 1832
Veronica Chamadrys
Wild Germander. Germander Speedwell
Diandria Monogynia

Plate 61

Skullcap ***Scutellaria galericulata***

Scutellaria Galericulata, Hooded Willow Herb,
Blue Skull Cap
Breamore Marsh, August 1834
Didynamia Gymnospermia 85

No. 653, Vol 4

Both English and scientific names refer to the calyx, which is shaped like a *galerum*, a leather skull-helmet worn by Roman soldiers.

Breamore Marsh Scutellaria Galericulata
Augst. 1834 Hooded Willow Herb. Blue Scull Cap
Dydynamia Gymnospermia

R5

Plate 62

Motherwort ***Leonurus cardiaca***

Leonurus Cardiaca, Motherwort
Hambledon, Surrey, July 1854
Didynamia Gymnospermia

No. 669, Vol 4

Motherwort is a continental species introduced to Britain as a medicinal herb in the Middle Ages, when it was given chiefly to ease pain in childbirth. It now occurs as a casual in widely scattered localities, particularly by old walls.

Leonurus Cardiaca
Motherwort
Didynamia Gymnospermia
Hambledon
Surrey
July. 1854

Plate 63

Bugle ***Ajuga reptans***

Ajuga Reptans, Common Bugle
Lyne Lane, Chertsey, Surrey, May 31, 1843
Didynamia Gymnospermia

No. 679, Vol 4

The plant on the right of the picture is one of the few in the entire collection not entirely finished by Caroline May.

Ajuga Reptans
Common Bugle
Didynamia Gymnospermia
Lyne Lane
Chertsey Surrey
May 31. 1843

Plate 64

Mezereon ***Daphne mezereum***

Daphne Mezereum, Mezereon, Spurge Olive
Stanwell, Middlesex, March 1834
Octandria Monogynia 75

No. 729, Vol 4

Always a scarce and retiring species (it was not officially 'discovered' in Great Britain until 1752) and probably only truly native in woods on chalk and limestone, in this Middlesex location it was almost certainly a garden escape or introduction.

Stanwell Middlesex
March 1834

Daphne Mezereum
Mezereon. Spurge Olive.
Octandria Monogynia

75

Plate 65

Spurge-laurel ***Daphne laureola***

Daphne Laureola, Spurge Laurel, Laurel Mezereon
Shrubbery, Breamore, April 1834
Octandria Monogynia 65

No. 730, Vol 4

Shrubbery
Brennore
April 1834

Daphne Laureola
Spurge Laurel: Laurel Mezereon
Octandria Monogynia

Plate 66

Birthwort ***Aristolochia clematitis***

Aristolochia Clematitis, Common Birth-wort
Oxfordshire, June 7, 1847
Hexandria Hexagynia

No. 731, Vol 4

Birthwort is a Mediterranean and Central European species which was introduced to physic gardens in the twelfth century. The shape of its extraordinary flowers reminded medieval herbalists of a uterus, so, according to the Doctrine of Signatures, it was given to women to speed childbirth, and to induce abortion (modern medical research has confirmed that it does have this property). The only remaining colonies in Britain are mostly naturalized on the site of old monasteries, and it is likely that the Oxfordshire plants which Caroline May painted were from the famous (and still surviving) colony near the ruins of Godstow Nunnery.

Aristolochia Clematitis
Common Birth=wort
Hexandria Hexagynia
Oxfordshire
June 7. 1847.

Plate 67

Box ***Buxus sempervirens***

Flowers and berries of Buxus Sempervirens,
Common Box-Tree
Box Hill, Surrey, September 1851
Tetrandria Trigynia

No. 742, Vol 4

This is one of the sites where box is almost certainly native. All the leaves, but not the flowers or seed, are varnished, to make the painting even more lifelike.

Flowers & Berries of
Buxus Sempervirens
Common Box-tree
Tetrandria Trigynia
Box Hill
Surrey Sep. 1851.

Plate 68

Small nettle ***Urtica urens***

Urtica Urens, Small Nettle
Rubbish, Hardwick Lane, October 17, 1842
Tetrandria Monogynia

No. 745, Vol 4

Urtica Urens
Small Nettle
Tetrandria Monogynia

Rubbish
Hardwick Lane
Oct 17. 1842.

Plate 69

Hop ***Humulus lupulus***

Male and Female Blossom of the Humulus Lupulus,
Common Hops
Garden, Breamore Rectory, September 1833
Pentandria Digynia 99

No. 747, Vol 4

Male & Female Blossom
of the
Humulus Lupulus
Common Hops
Pentandria Digynia
99
Garden
Rectory
Sep 1833

Plate 70

English elm ***Ulmus procera***

Ulmus Campestris, Common Elm
Ruxbury, Surrey, March 1847
Pentandria Digynia

No. 749, Vol 4

Probably the English elm, but elms are a very variable family, and the specific name *U. campestris* was confusingly used for several different species.

Ulmus Campestris
Common Elm
Pentandria Digynia
Surrey
March. 1847

Plate 71

Willow species possibly ***Salix purpurea***

Near the Mills, Chertsey, March 1847

No. 762, Vol 4

near the Mills
Chertsey.
March 1847

Plate 72

Juniper ***Juniperus communis***

Juniperus Communis, Common Juniper
Ruxbury, Chertsey, May 19, 1848,
Found wild on Breamore Downs, Hants
Monadelphia Triandria

No. 780, Vol 4

The fruit was almost certainly added later, but is undated.

Juniperus Communis
Common Juniper
Monadelphia Triandria

Buxbury
Chertsey
May 19. 1848.
found wild on Breamou
Downs. Hants —

Plate 73

Bee orchid ***Ophrys apifera***

Ophrys Apifera, Bee Orchis or Twayblade, Bee Flower
Down Fields, Breamore, July 18, 1837
Diandria Monogynia

No. 815, Vol 5

The name 'twayblade' usually refers to a quite different species, *Listera ovata*.

Down Fields.
Breamore
July. 18. 1837.

Plate 74

Stinking iris ***Iris foetidissima***

Iris Foetidissima, Stinking Gladdon or Flag
Lane near Teignmouth, June 1831
Triandria Monogynia

No. 818, Vol 5

This species gets its malodorous names from the leaves, which smell slightly of stale beef when crushed. But they hardly merit the adjective 'stinking', and a pleasanter and more appropriate name is roast-beef plant. 'Gladdon' comes from the Old English *glaedene*, meaning 'a little sword'.

Iris Fœtidissima
Stinking Gladdon, or Flag
Triandria Monogynia
Lanes near
Teignmouth
June 1831.

Plate 75

Widow iris ***Hermodactylus tuberosus***

Iris Tuberosa, Snake Iris
Bitton, Gloucestershire, April 1868
Florideae Iridaceae

No. 819, Vol 5

A Mediterranean species, grown in gardens, and naturalized in a few places in the south-west of Britain.

Iris Tuberosa

Snake Iris

Floridae Iridaceæ

Bitton

Gloucestershire

April 1868

Plate 76

Butcher's-broom ***Ruscus aculeatus***

Ruscus Aculeatus, Knee Holly, Butchers Broom
Breamore Wood, March 1835
Triandria Monogynia 103

No. 830, Vol 5

This curious member of the lily family was so called because the spiny shoots were used for scouring butchers' blocks.

Breamore Wood
March 1835

Ruscus Aculeatus
Knee Holly Butchers Broom
Triandria Monogynia

Plate 77

Fritillary ***Fritillaria meleagris***

Fritillaria Meleagris, Fritillary,
Chequered Daffodil, Snake's-Head
From a specimen in a garden at Chertsey. Found wild
in Middlesex and several other counties, April 5th, 1841
Hexandria Monogynia

No. 831, Vol 5

Before 1930 the fritillary occurred in at least twenty-seven different counties of Britain. Its favoured habitat was hay-meadows subject to winter flooding, and modern drainage and grassland conversion have reduced it to a score of individual meadows. In nineteenth-century Middlesex there were fritillary meadows at Iver, Swakeleys, Ruislip Common, Mortlake and Pinner. Now there are none.

Fritillaria Meleagris
Fritillary. Chequered Daffodil. Snake's=head
Hexandria Monogynia
From a specimen
in a garden at Chertsey
found wild in Middlesex
& several other counties
April 5th 1841.

Plate 78

Black sedge ***Carex nigra***

Carex Caespitosa, Tufted Bog Carex
Pool at Siston Common, Kingswood, Gloucestershire, June 1859
Triandria Monogynia

No. 880, Vol 5

Carex Cespitosa
Tufted Bog Carex
Triandria Monogynia
Pool at Siston Common
Kingswood Gloucestshre
June 1859

Plate 79

Long-bracted sedge ***Carex extensa***

Carex Extensa, Long Bracteated Carex
Braye du Valle, Guernsey, July 1862
Monoecia Triandria

No. 884, Vol 5

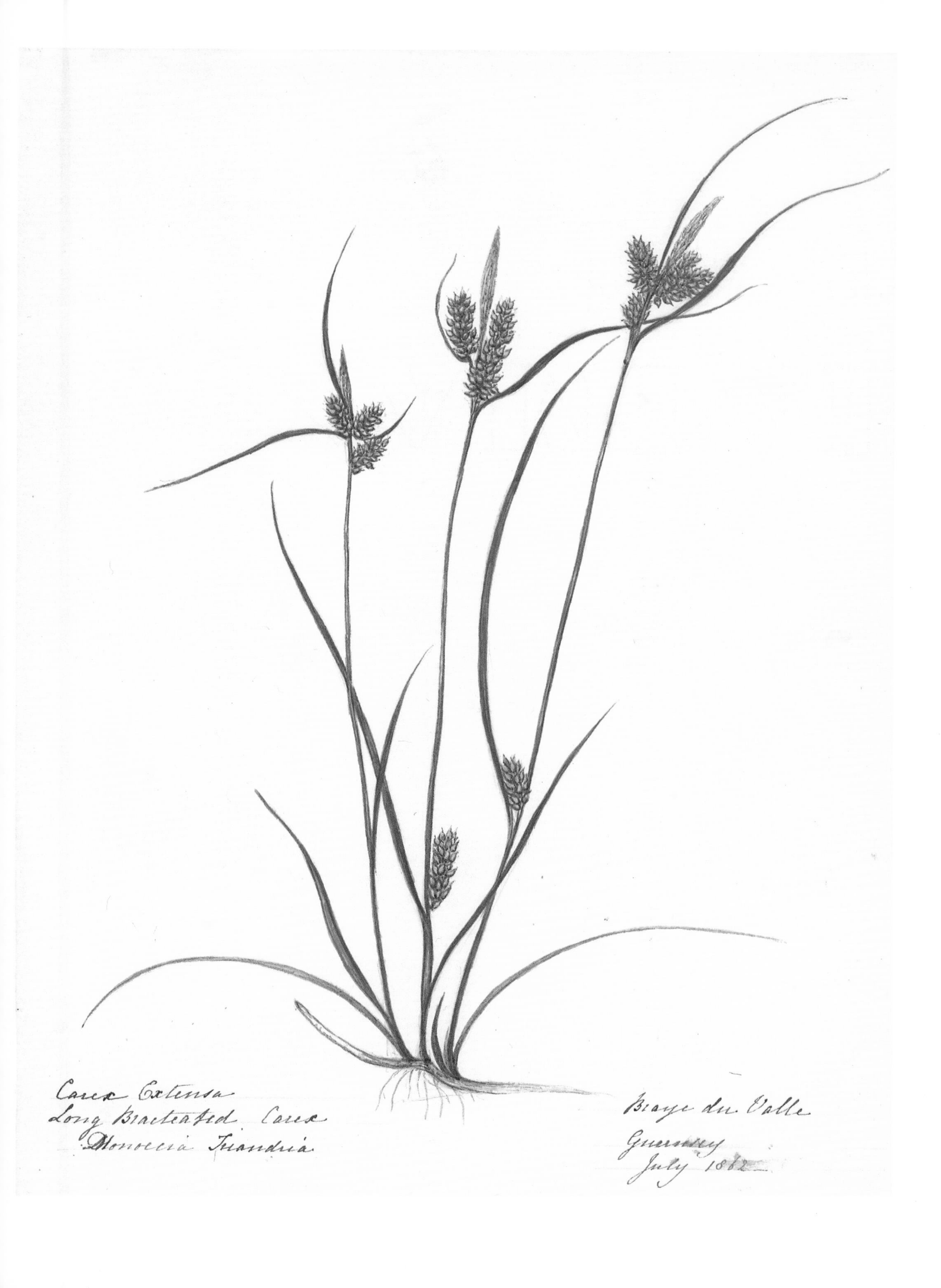
Carex Extensa
Long Bracteated Carex
Monoecia Triandria
Braye du Valle
Guernsey
July 1862

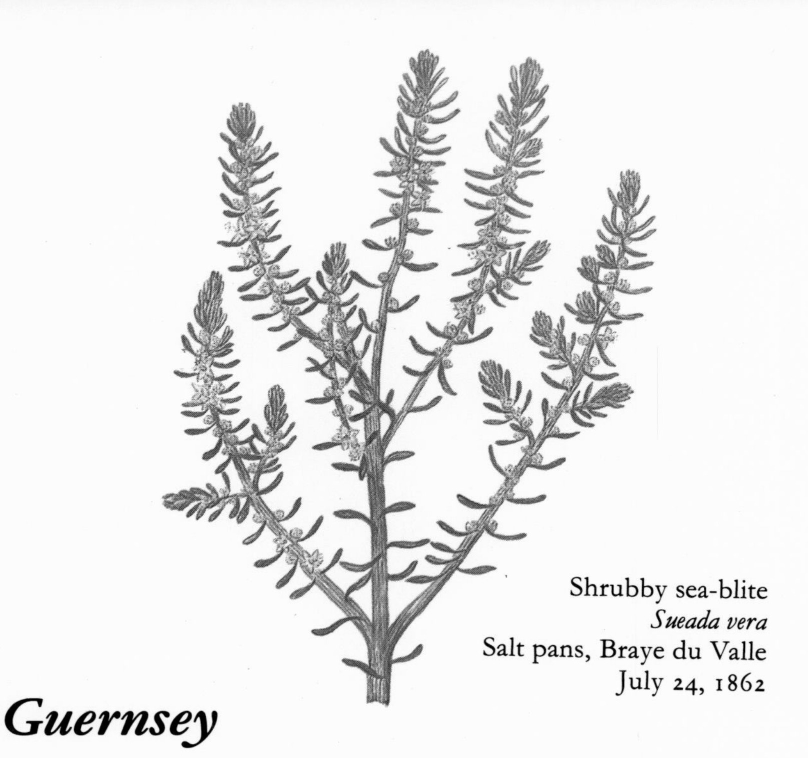

Shrubby sea-blite
Sueada vera
Salt pans, Braye du Valle
July 24, 1862

Guernsey

The crossing by paddle steamer from Bridport would not have been undertaken lightly when Caroline made her first visit there in 1858. We do not know if she travelled alone, with her sisters, or where she stayed, but as an energetic spinster of 49, she certainly got about the island. Eighty-one paintings from the complete flora come from Guernsey, from sites stretching from Fort Doyle in the north to Torteval in the south-west. She must also have travelled to the nearby island of Sark, to paint four of the plants there.

She could hardly have failed to notice the spectacular scenery of this beautiful island on her four extended visits, but her efforts were concentrated, as ever, closer to the ground. The botany of Guernsey was of interest to Caroline, as to scientists before and since. Barbet's map of 1869, shown here, includes a scattering of small letters, indicating the presence of wild flowers, named in a key on the original map. The letter 'b' indicates *Polypogon monspeliensis* in the Braye du Valle (ringed), which is indeed where Caroline found it in September 1860 (see Plate 81). She must have been especially interested in this area, only reclaimed from the sea in 1803, and in the Landes and L'Ancresse areas where great stretches of sand are held together by grasses, gorse, brambles and bracken.

Among her finds were several 'first recordings', including the shrubby seablite *Suaeda vera*, shown above, which she painted in 1862. She also painted, in 1858, the pretty lesser water-plantain, *Baldellia ranunculoides*, below. It must have been rare then. It is now extinct in Guernsey.

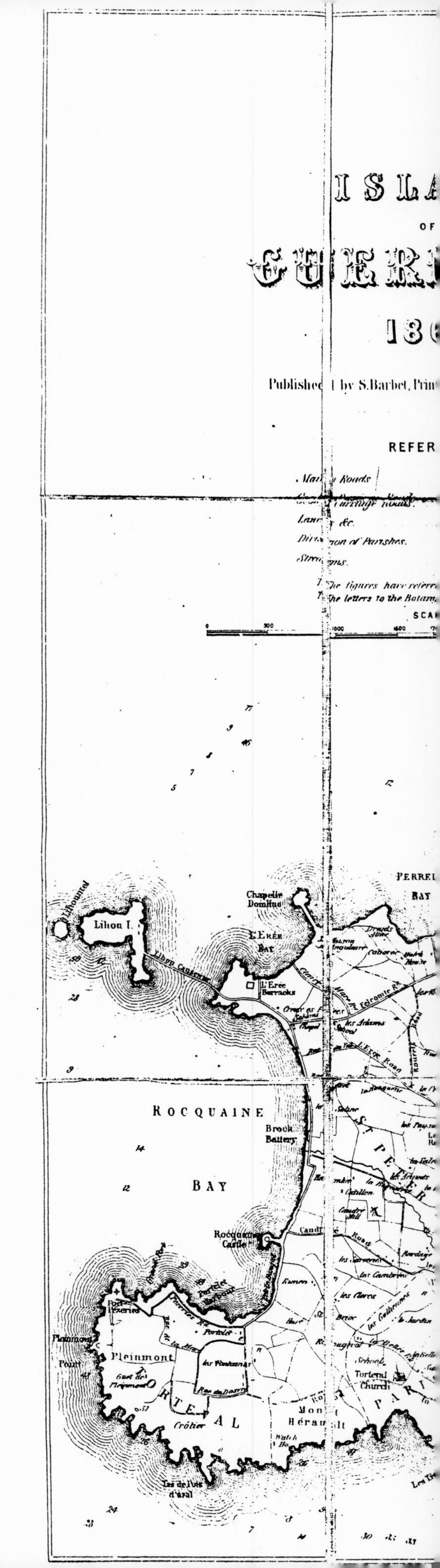

Lesser water-plantain
Baldellia ranunculoides
Lancresse Common
May 1858

Fort Le Marchant
Fort Doyle
LANCRESSE BAY
Grand Camp
Mont Cuet
Tower
Druids Altar
Les Grais
GRAND
HAVRE
Lancresse Common
Vale Church
Port Grat
Piquerel
Pequeries
Portinfer
Battery
Port Soif
Grandes Rocques
Long Isle Road
BRAYE DU VALLE
Braye Road
Vale Castle
Bordeaux Harbour
Mont Crevelt
Saline B.
Videclin
Long Port
COBO BAY
Cobo Road
Hommet
Burton Batt
St. Matthews Church
Les Genats Road
Petites Capelles
Capelles School
Chapel
Hougue a la Perre Battery
Belle Greve Batt
Long Store
Salerie Battery
White Rock Landings
NEW HARBOUR
Castle Cornet
TOWN OF ST PETER PORT
Victoria Tower
St Johns Church
Saumarez Road
CATEL PARISH
Catel Church
LA GRANDE MARE
Kings Mills
ST PETER PORT
Les Terres
Gentlemens Bathing Places
Ladies Bathing Places
Clarence Batt
Petit Fort Bay
FORT GEORGE
Kent Battery
Fermain Bay
St Martins Church
ST MARTINS PARISH
Bec du Nez
Doyles Pillar
Jerbourg Barracks
JERBOURG
St Martins Point
Moulin Huet
Saints Bay or Baie de Seing
Icart Point
Icart Road
Tower
Mont Hubert
Petit Bot Bay
Sommeilleuse
Moye Point
Le Gouffre
FOREST PARISH
Forest Church
Forest Road
St. Saviours Church
Torteval
Bon Repos

Plate 80

(?) **Greater pond sedge** ***Carex riparia***

Carex Stellulata, Little Prickly Carex, May 1856
Triandria Monogynia

No. 895, Vol 5

There is some doubt about the identity of this species. *Carex stellulata* is the obsolete synonym for *C. echinata*, which is very different from the species here. The illustration most closely resembles the greater pond sedge, *C. riparia*.

Carex Stellulata
little prickly Carex
Triandria Monogynia

May. 1856

Plate 81

Annual beard-grass ***Polypogon monspeliensis***

Polypogon Monsepeliensis, Annual Beard Grass
Braye du Valle, Guernsey, September 20, 1860
Triandria Monogynia

No. 903, Vol 5

A rare species of damp coastal grassland in the south of Britain.

Polypogon Monspeliens
Annual Beard Grass
Triandria Monogynia
Braye du Valle
Guernsey
Sep. 20. 1860

Plate 82

Common reed ***Phragmites australis***

Arundo Phragmites, Common Reed
Near Les Landes Vale, Guernsey, September 17, 1860
Triandria Digynia

No. 907, Vol 5

Arundo Phragmites
Common Reed
Triandria Digynia
near Les Landes
Vale
Guernsey
Sep. 17. 1860

Plate 83

Bird cherry ***Prunus padus***

Prunus Padus, Bird Cherry
Field at Cartha Martha, Cornwall, May 1873
Icosandria Monogynia

No. 949, Vol 5

This painting is the last in this selection to be painted by Caroline May, in the year before she died. It, and the painting which follows, are part of a miscellaneous group placed at the end of Volume 5 and painted in the last three years of her life. It seems that she could not face the time-consuming rearrangement of the albums to preserve her taxonomic order.

Prunus Padus
Bird Cherry
Icosandria Monogymia

Fidd at
Carthia Martha
Cornwall
May 1873.

Plate 84

Medlar ***Mespilus germanica***

Mespilus Germanica, Common Medlar
Copse in hedge, Trevosa, Sth Petherwyn, Cornwall, June 1871
Nat. Order Calyciflorae Rosacea
Icosandria Pentagynia

No. 967, Vol 5

This southern European tree is probably not native in Britain, but in the south it occasionally becomes naturalized from the bird-sown seed of garden specimens.

Mespilus Germanica
Common Medlar
Icosandria. Pentagynia
Nat Order Calyciflorae Rosacea
Copse in hedge
Trevosa
Sth Petherwyn Cornwall
June 1871

Acknowledgements

The publishers, Richard Mabey and John Tyler would like to express their sincere thanks to Grace Tyler, née May, for her encouragement, and to all those who have helped with illustrations and information for the text sections of this book, and particularly to David Bowyer, Chertsey Museum, Patrick Hickman, Lieselotte Killick, Michael Martyn, David McLintock, Bernard Pardoe, Priaulx Library (St Peter Port Guernsey), Rev. Gordon Robinson, Surrey Local Studies Library (Guildford), Garry Strutt, Paul Tyler and Sir John Wordie.

The map on page 35 is courtesy the Hampshire Record Office, Winchester; on page 111 courtesy the Cornish Studies Library, Redruth; and on page 181 courtesy the Priaulx Library, Guernsey.

The engraving of Botleys Park on page 84 is by G.F. Prosser, from his *Select Illustrations of Surrey*, 1828